D0410063

JON FINCH
BEER
CRAFT
THE NO-NONSENSE GUIDE
TO MAKING AND ENJOYING
DAMN GOOD CRAFT BEER
AT HOME

sphere

JOKER
Welsh Red Ale
Tiny Rebel
Cali
AMERICAN
PALE ALE

338
339

CONTENTS

FOREWORD

BY JAMES SPENCER
HOST OF BASIC BREWING PODCASTS

IF YOU LOVE GOOD COMMERCIAL BEER, YOU CAN THANK HOME BREWING. THE CRAFT BEER MOVEMENT WAS BORN IN THE STEAM RISING FROM COUNTLESS KETTLES ON HOME COOKTOPS AND NURTURED BY THE PRAISE OF FRIENDS AND NEIGHBOURS AS THEY SAMPLED WONDERFUL BREWS.

IN THE OLD DAYS, HOME BREWING SUFFERED FROM A BAD REPUTATION. FERMENTING CANNED EXTRACT WITH LOTS OF TABLE SUGAR IN OPEN CROCKS OFTEN LED TO CONTAMINATED BEER AND EXPLODING BOTTLES. TODAY, HOME-BREWING TECHNIQUES ARE VERY SOPHISTICATED, AND HOME-BREWERS HAVE ACCESS TO THE SAME HIGH-QUALITY INGREDIENTS THAT PROFESSIONALS USE. WE ARE LIMITED ONLY BY OUR IMAGINATIONS.

THERE ARE MANY APPROACHES TO BREWING BEER AT HOME. SOME PEOPLE BREW SMALL VOLUMES OF BEER WITH SIMPLE EQUIPMENT, WHILE OTHERS FILL THEIR GARAGES AND BASEMENTS WITH HIGH-TECH GADGETRY AND CRANK OUT LOTS OF DELICIOUS BREW. SOME STICK TO BASIC STYLES, WHILE OTHERS SEEK TO BREAK TRADITIONAL BOUNDARIES AND BREW THEIR OWN CREATIONS.

WHETHER YOU'RE AN ASPIRING COMMERCIAL BREWER OR SOMEONE WHO WANTS TO STICK CLOSER TO HOME, BREWING BEER CAN BE A REWARDING CREATIVE EXPERIENCE. ONE THING'S FOR SURE: YOU'LL NEVER LOOK AT BEER THE SAME WAY AGAIN.

WHAT IS THIS BOOK?

FACT: IF YOU CAN MAKE SOUP, YOU CAN MAKE BEAUTIFUL, TASTY AND DELICIOUS FRESH CRAFT BEER.

I WROTE THIS BOOK TO HELP YOU BREW AWESOME BEER.

IF YOU LIKE COOKING AND YOU LIKE BEER, THEN GIVE HOME BREWING A GO. IT'S A MILLION MILES OFF THE 'HOME-BREW KITS' FROM YEARS AGO THAT GENERALLY PRODUCED FOUL-SMELLING, CLOUDY, YEASTY BREWS. ARMED WITH THIS BOOK, A STOCK POT AND A MESH BAG, YOU'LL BE DRINKING YOUR OWN BEAUTIFULLY CRAFTED, FRESH, HOPPY AROMATIC BEERS IN A MATTER OF WEEKS. I PROMISE.

I DO NOT USE BAFFLING TERMINOLOGY, DESCRIBE WEIRD, LONG-WINDED PROCESSES OR TELL READERS THEY NEED TO GO AND BUY EXPENSIVE, BULKY EQUIPMENT. THIS IS NOT A CHEMISTRY TEXT-BOOK WRITTEN BY A SCIENTIST.

I'M A LAZY BEER GUY.

I APPROACH HOME BREWING THE OPPOSITE WAY, BY TAKING ALL THE NECESSARY SCIENCE AND THE GEEKINESS BUT PUTTING IT IN LAYMAN'S TERMS. I STRIP OUT ALL THE UNNECESSARY COMPLEXITY IN THE PROCESSES AND EQUIPMENT (THAT YOU JUST DON'T NEED), MAKING BREWING BEER FUN AND APPROACHABLE.

I HAVE NOT INTENDED IN THIS BOOK TO EXPLAIN ANY OF THE JARGON OR TECHNICALITIES OF BREWING, NOR IS IT INTENDED AS A SPECIALIST GUIDE TO DEVELOPING OR FOLLOWING HOME-BREW RECIPES. THERE ARE HUNDREDS OF BOOKS OUT THERE ALREADY THAT DO A FAR BETTER JOB OF THAT THAN I EVER COULD.

HOPEFULLY IT'S ALL CLEAR ENOUGH TO GIVE YOU THE CONFIDENCE JUST TO GRAB A STOCK POT FROM THE KITCHEN CUPBOARD AND MAKE SOME GREAT CRAFT BEER IN YOUR OWN HOME.

IT'S SIMPLE, CHEAP AND EASY TO DO. I BREW BECAUSE I REALLY ENJOY THE PROCESS AND I REALLY LOVE THE RESULTING BEERS! I HOPE YOU DO, TOO.

WELCOME TO YOUR NEW FAVOURITE HOBBY.

HOW TO BREW BEER

HOW TO BREW BEER

I use a method called 'Brew-in-a-Bag' that was developed and became popular in Australia. It is a process of brewing beer with malted grains of barley (all-grain) without the need for the expensive and complex equipment most serious home brewers use. Rather than using large brewing vessels, all you need for this method is a big pan and a mesh bag.

When I started brewing, my first few batches were made in a pasta pan that I had to hand in the kitchen. The only extra kit I needed to buy to brew beer was a mesh straining bag and a large plastic bucket to ferment in. I've since upgraded and have a dedicated 33L brewpot just for brewing beer but, let me tell you, that smaller pan produced equally good beer!

I've added a kit list later in the chapter. This isn't by any means an exhaustive list – it's just the bare minimum I recommend you need to get brewing at home.

SIMPLY PUT, THE BASIC STEPS TO BREWING BEER MY WAY ARE:

USE A MESH BAG TO STEEP MALTED BARLEY GRAINS IN HOT WATER SO THEY RELEASE THEIR SUGARS

DRAIN AND THEN BOIL THE RESULTING SUGARY WATER FOR AN HOUR OR SO

ADD HOPS AT VARIOUS STAGES FOR FLAVOUR AND AROMA

COOL THE LIQUID DOWN, POUR INTO A LARGE BUCKET AND ADD YEAST TO FERMENT

CARBONATE, DRINK AND ENJOY THE END RESULT

REPEAT

MALTS

HERE'S A FACT THAT WILL BLOW YOUR MIND APART. VIRTUALLY EVERY BEER YOU DRINK STARTS OFF THE SAME, WITH 80–90% IDENTICAL MALTED BARLEY AS THE BASE.

The purpose of this base malt is to provide the fermentable sugar we need to make the base beer – then we add the speciality malts to really boost interest and personality. Speciality malts are not there to provide sugars; we're adding them to give interesting colour and flavour to the beer. Think of it like a cake. Most cakes are made from flour, but it's the addition of chocolate that makes it a chocolate cake, the addition of cherries that makes it a cherry cake, and so on. In beer terms, the base malt is the flour and it's what we add to the mix that determines what kind of beer we'll get. Besides the speciality malts, other things affect what beer we'll end up with – the hops, our yeast choice, or weird and wonderful additions like fruit, herbs, botanicals etc.

Malted grain is made by sprouting barley grains until they germinate and then quickly drying them in a kiln. The longer they are roasted in the kiln, the darker the grain and generally the more intense the flavour becomes, going from a light bready colour and aroma through to darker caramel flavours and then to a rich, dark, black burnt bitterness. Different varieties of grain also give different complexities and characteristics.

Most beers are made using a number of different kinds of malted barley, each lending its own flavour, colour and characteristic to the beer. You'll see in the recipes later in the book that virtually all of them call for 4–5kg of base malt and then an additional 500g or so of other malts depending on the beer style.

So a thick dark pint of stout is mainly made using pretty much the same grains as a pilsner. What differentiates the two is that stout also contains some darker, roasted malt to give it that lovely colour and roasted coffee flavour. Not much, though – we're talking perhaps just 10% of all the grain used.

When you are looking to buy malts, you'll usually see a number next to each one. This may be called SRM, EBC or Lovibond. Don't worry about the terminology – all you need to know is the higher the number, the darker the malt and therefore the darker it is going to make your beer.

A 20 EBC malt, for example, will be much lighter and mellower than a 150 EBC malt.

BREWING WITH MALT EXTRACT

Years ago, making beer from kits was the big thing.

You'd get a large, heavy tin of gloopy malt extract, dump it in a bucket, add your water and chuck in some yeast.

It never made good beer because the quality of the main component, the malt extract, was terrible.

Happily for us now, times have changed. Malt extract producers have raised their game a long way and you can buy top-quality products right off the shelf. Extract comes in two different forms – dried and liquid.

The first step in the beer-making process is to get these lovely malted grains that are full of sugar and steep them in hot water for a while so all the sugars are released. Then we bring the liquid to a boil and continue the beer-making process.

To make malt extract, the manufacturers steep the grains the same way we do to get the sugary liquid and then they boil as much water off it as they can to leave a sticky, intensely sweet syrup. This is then either canned as it is, or spray-dried to make dry malt extract powder.

To use malt extract in your brewing, just add it to your water and then boil it up with hops the same way you would if you'd steeped the grains yourself. So the important thing is that the beer-making process is identical from that point onwards. All malt extract is doing is speeding up the time it takes to get your sugary liquid ready to boil. It's skipping the work of steeping the grains for an hour or two. Saves you time and effort. It also saves hoisting a big, heavy sack of hot, dripping grains around your kitchen.

You may ask why we shouldn't just use malt extract for every single beer as it sounds much easier. Well, first up it is more expensive. The cost of malt extract needed to make an average 20L batch of beer would be around £15 whereas I can get the same result from using around £4.50 worth of malted grain.

Cost aside, the other main reason is creativity. Using a malt extract means you are using whatever grain the manufacturer has chosen. It doesn't give you a great deal of scope to play around with a recipe, but by using all grain you can tweak and dial in the recipe as you see fit.

Using extracts used to be seen a bit as cheating but it's really not. Especially with the modern production processes.

There is a middle ground where the base beer is made from malt extract and then speciality grains are steeped in hot water separately and added to the pot. This is a very popular option as it's easy to do and allows for versatility.

HOPS

Latin name: *Humulus Lupulus*, literally meaning 'small wolf' or 'wolf of the woods'. They belong to the Cannabaceae family, and are distantly related to the cannabis plant.

Worldwide there are around 250 varieties of hops. The largest producer of hops is the USA, with just under 22,000 hectares, closely followed by Germany, with around 18,500 hectares. In comparison, the UK has only 910 hectares.

Generally speaking, hops fall into two categories: aroma hops and bittering hops. Some varieties will be used only for bittering and some only for their aroma, while others are dual-purpose.

Currently, the most sought-after hop characteristics from brewers include grapefruit, citrus and tropical fruit flavours, but other aromas such as floral, herbal, spicy, pine, black pepper, grassy, coconut, bubblegum and blackcurrant can also be found – and many, many more besides.

It's crazy to think such a diverse mix exists but this is the very thing that has made the recent craft-beer revolution so exciting. The range of flavours and aromas professional and home brewers alike have at their fingertips is incredible.

Before being used as an ingredient in beer, hops were used in early Egyptian times as a medicinal herb. They contain an antibacterial agent and can have anti-inflammatory and anti-spasmodic effects, and as such have been used to treat skin ulcers, liver disease and digestive problems.

Reading between the lines here it is safe to argue that hops should be seen as one of nature's medicines. Doctor's orders, if you will. And based on that, in my head anyway, it's not much of a leap to claim that beer is good for you. It's science, so you can't argue.

THE UK HOP YEAR

MARCH

The hop 'yard' or 'garden' (depending on where in the country you are) is strung ready for the young bines to climb. This is done by hand using a 'monkey' – a long pole with a hook at the end. Some growers will, however complete this task over the winter months.

APRIL/MAY

First shoots emerge from the soil and are trained clockwise up the strings by hand. Dependent on the variety, two or three shoots are tied while the rest are removed either chemically or mechanically.

JULY/AUGUST

By mid to late July the hops will have reached their full growing height and will begin to shoot laterals. The hops come in to burr for around three weeks before coming in to hop.

SEPTEMBER

The hop bines are harvested in the field and transported back to picking sheds. Here the hop cones are stripped mechanically from the bine before being kiln-dried to reduce moisture content from 80% down to only 10%. The dried hops are then pressed into pockets or bales and transported to the hop merchant for further processing.

OCTOBER/NOVEMBER

Once picking is complete, the bines are cut to the ground and disposed of.

YEAST

PROBABLY THE SINGLE MOST IMPORTANT INGREDIENT IN BEER IS YEAST.

Amazingly, the strain of brewer's yeast used can have as much – if not more – impact on the flavour and character of your beer than any other ingredient.

Some yeasts are designed to allow the beer to finish crisp and dry with very little yeast character. This generally allows the flavours of the hops or the malts to shine through. A good example would be an American pale ale which would give you the taste of big hops with a clean finish.

Other yeasts are designed to really impart flavours like the funky barnyard aromas and flavours in a saison, or the clovey, often banana-like aromas found in German wheat beers. There's no banana or clove in the recipe – those flavours come from the yeast.

In short, the beer style you are brewing will usually dictate the strain of yeast you choose. Wheat beers, lagers, pale ales, London ales, Scottish ales, Irish ales, bitters, west coast, Belgian triple etc – they all have an appropriate strain to use.

Many home brewers choose to rehydrate their yeast sachets before they use them – it is said to give the yeast a head start. Frankly, as I am lazy I have never done that and have never had an issue. The important thing is to use FRESH sachets or vials that you purchased recently. Don't use something that has been sitting in your shed for a year.

One way to experiment with your home brewing is to try unusual or creative combinations of beer style with yeast choices. So, choose a lager recipe but then brew it with an Irish ale yeast or a saison yeast, for example.

Hundreds of strains are readily available online both in dehydrated sachet form or in a liquid vial.

FERMENTING

This is where the magic happens. Where your sweet, sticky concoction turns into delicious beer. It amazes me every time.

Once the yeast is added and your fermenter (the large bucket used for fermenting) sealed, place it somewhere that has a constant temperature of between 18 and 22 degrees Celsius. Yeast can handle being slightly outside this range but it dislikes the temperature moving around too much. Consistency is key, and err on the side of cooler over warmer if you can.

You'll notice your airlock (which acts like a simple one-way valve that allows the carbon dioxide to escape without allowing nasties to enter your beer) bubbling away after about 24 hours and this will be vigorous for two to three days before it slows down and eventually stops.

Many older home-brew guides will advise moving your beer off the 'dead yeast' at this point but we're not going to do that. In fact my recommendation is do not even lift the lid of your fermenter for a peep until a full two weeks have passed.

Tests carried out show that moving beer any earlier does not improve the final product whatsoever, yet risks exposing your precious golden beauty to nasty bugs. By the time it is two weeks old, it's a little more robust and resilient, and also most of the sediment and yeast will have dropped to the bottom of the fermenter, meaning you can draw off cleaner, brighter beer.

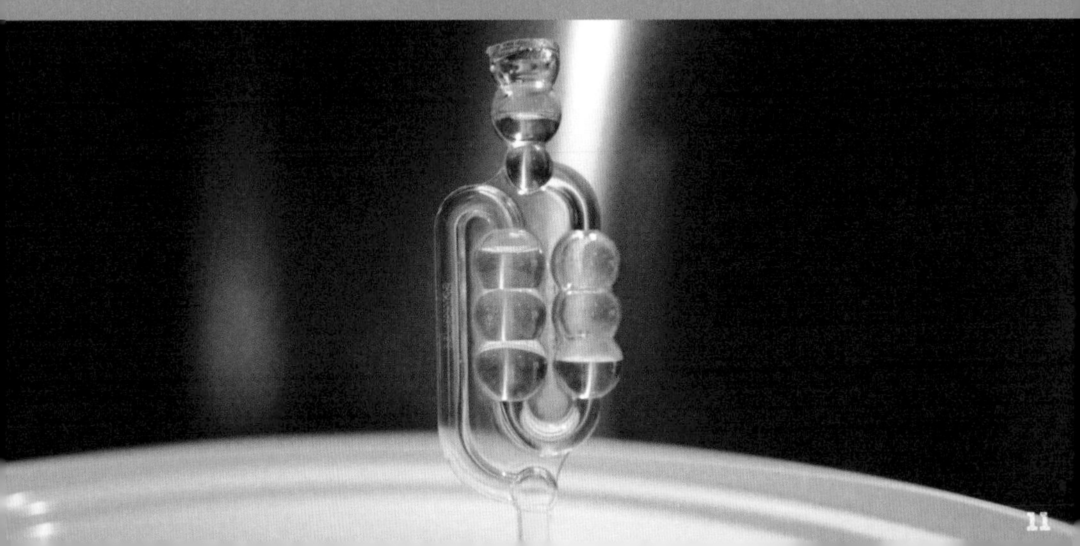

CONDITIONING

Once your beer has fermented, technically you can drink it straight away. What you'll find, though, is it doesn't taste all that great. It will still have residual yeast in there and all the flavours will taste like they're clashing. The beer needs a couple of weeks just to mellow out. To calm down and let everything find its place. All the flavours should then come together in a glass of harmonious liquid perfection.

Back in the day, brewers would move their beer to a secondary bucket specifically to condition it, but there's really no need. Once it has had a couple of weeks in the fermenter it's time to move it to either a keg or bottles, where it will happily sit for another week or two.

THEN you can drink it.

A good rule of thumb is for every percentage of alcohol over 2%, that's how many weeks you might condition your beer before drinking. So if your beer is around 5%, best to give it three weeks. Big, high-strength beers like imperial stouts or barley wines might need a good couple of months.

That's just a broad-brush approach, though, as certain beer styles drink better with longer or shorter ageing times. For example, a hefeweizen/wheat beer is typically drunk straight away. Literally as soon as it is finished fermenting you should get stuck in. A pale ale with tons of hops should also be drunk fairly fresh – within a month at the least – as you'll start losing the delicate aromas and flavours that make this beer the superstar that we all love. Other beers like strong IPAs and festive beers might benefit from a good three to four months.

The thing is, you can just keep cracking open a bottle or pouring yourself a glass from the fermenter every few days until the beer is how you like it. That's one of the great things about home brew: you are brewing for yourself. Brew what you like and drink it when it tastes good to you. No rules here.

DRY HOPPING

Ten years ago, dry hopping was something you did in a dark corner of a dingy night club in Rotherham.

But actually, there is barely a single modern brewer out there who has not embraced this technique. It's how brewers get that massive hit of hops. That smack-in-the-face citrus and knock-out smell that we all love about pale ales and IPAs. The trick is to allow the beer to ferment for a few days and then dump a lot more hops right into the fermenter and leave them there a week or so. Like a beery hop tea. All the big breweries use this technique to make those big hop-bombs too.

This method extracts all the lovely, but volatile, aromatics from the hops that would otherwise be boiled away in the hot liquid. Hops have natural antibacterial properties so providing you are using them straight from the pack there's no real worries about contaminating your beer. Also, at this point your beer (hopefully) contains a useful bit of alcohol that will also kill off any unwanted nasties.

There are a few ways to dry hop, but here's how I do it.

Buy yourself a pack of women's knee-high tights. I opt for plain as I'm not sure whether the coloured ones would leach dye into my beer or not. Despite the rumours, I have not experimented all that much with women's undergarments. Anyway, back to the dry hopping . . .

For an average 20L batch of beer, I'd opt for a good 50g of dry hops. And make sure you are using big American-aroma-style hops, not hops that are designed purely to add bitterness to beer. Stuff them into the tights along with a small handful of sanitised marbles (to help weigh the sack down). Tie at the top and gently submerge it in your beer. You can attach some fishing wire to the top for easy removal if you like.

I've also had great success with chucking a sack of hops into a keg before filling it with freshly fermented beer and then just leaving it in there as I drink my way through the keg. You'd think the beer would end up overly hoppy or contaminated in some way but I've never found that. It goes against the advice you'll read elsewhere, but hey.

When you are done, open the sack, retrieve your marbles then sling the rest in the bin. That's why I use knee-high tights because washing out and cleaning purpose-made hop bags sucks.

BOTTLE OR KEG?

BEER NEEDS PACKAGING IN ORDER TO CARBONATE IT, STORE IT AND SERVE IT. THERE ARE THREE MAIN OPTIONS OPEN TO HOME-BREWERS AND THEY ARE:

- **BOTTLING**
- **PLASTIC KEG, E.G. KING KEGS**
- **CORNELIUS KEGS**

The choice comes down to a number of factors such as your budget, how much free time you have, whether you wish to transport your beer regularly, storage requirements, serving requirements etc.

Most home brewers start with bottling. It's cheap (you can usually get bottles for free from the local pub or by asking friends to save their empties) so the only kit needed is a little bottle-capping tool and some caps. It's easy to transport beer to friends' houses and it can be easily stored in the fridge, basement, garage or spare cupboard for short or long periods of time. However, it is time-consuming, with each bottle needing to be individually washed, rinsed, sanitised, filled and capped. The beer must also be primed with a sugar solution or drops to carbonate the beer.

Plastic kegs are a logical option for those with no wish to transport their beer. It is one vessel to clean and fill, which takes a fraction of the time bottling takes. As with bottling, the beer in the keg must be primed to ensure carbonation. The main downside to king kegs is the lack of control over carbonation. As the keg starts to empty, the carbonation level will drop, which in turn means flat beer. It also means dispensing becomes an issue. Some plastic kegs have the option of a special gas inlet valve that allows the CO2 to be topped up as required to ensure consistent carbonation levels. Note – this will not carbonate the beer, just maintain existing levels.

Although more costly, the best set-up in my view is to invest in a Cornelius ('corny') keg or two. These were originally made for soft drinks companies for dispensing their products in pubs, and many still bear the name of the original owner. While they are available brand new, they are very expensive so most people buy them reconditioned. Mine are actually classified as 'budget' kegs as they look a bit battered and dented. However, they are airtight and function just as well as a brand-new keg.

The advantages of corny kegs are numerous but the key feature for me is the ease of use. They hook up to a CO2 cylinder in the same way as pub beer taps, giving ultimate control over the gas levels from carbonation through to serving. And providing you follow some simple processes prior to kegging, they pretty much guarantee bright, sediment-free beer right down to the very last pint. Being airtight and lightproof they are also very good for conditioning beers for many months if required. Even better, they allow you to 'force carbonate' your beer in a matter of hours so you can be drinking it sooner. I recently carbonated my beer fresh from the conditioning bucket in just under two hours with a few tips and tricks.

Essentially, though, corny kegs are a home draft beer system.

The downside is they tend to multiply – my initial keg has now turned into four kegs, each connected to the master gas cylinder with a different beer on tap. I'm also planning a 'kegerator' build soon – a conversion of a fridge that allows you to store the corny keg inside with the tap mounted on the door allowing beer to be served at the perfect temperature all year round. And cleaning them out is a doddle.

I approached packaging my beer the opposite way to most. As I've said before, I'm quite a lazy brewer and knew that if I had to bottle my beer after each batch was ready then the enthusiasm for my new hobby would soon falter. I needed to make life as simple as possible so I went straight to plastic kegs and then within a couple of months to corny kegs. I don't often transport my beer, I have a shed with plenty of space to store the kegs and I like the convenience of having my various beers on tap.

I decided to give some of my 'Bad Santa' (see page 63) away to friends and family as Christmas gifts and so found the need to bottle some of my beer. Tools like a bottling wand make this task much easier but it's perfectly fine to do it with a simple siphon tube.

I'm over-cautious on sanitisation so my bottle-cleaning process is probably overkill.

Here it is:

- rinse bottles out with hot water, removing any visible crud with a bottle brush
- soak for an hour in Oxi-Clean
- rinse thoroughly
- load into empty dishwasher and run on hot cycle (no detergent added)
- blast through with Star-San
- fill
- cap, having first boiled the caps in water for ten minutes

In order to get your beer fizzy, you need to give the residual yeast a little sugary snack. You can mix brewing sugar and boiling water together then pour into your bucket to pre-mix, or simply buy carbonation drops and add one per bottle before you put the cap on.

You'll need to leave your bottled beer somewhere warm for a few days to restart the yeast and then transfer it somewhere a little cooler for another fortnight for the carbonation process to finish.

TAP EAST
STRATFORD
CITY

CLEANLINESS IS NEXT TO GODLINESS

SANITISATION

THERE IS NO SINGLE MORE IMPORTANT ASPECT TO BREWING GOOD BEER THAN ENSURING GOOD SANITISATION.

READ THAT LINE AGAIN UNTIL IT SINKS IN.

Regardless of how fancy your hops are and how perfectly crushed your grain is, if your kit is not clean and sanitised, your beer will suck.

If you are complacent about this at any stage of the brewing process it is likely your beer will become contaminated or infected. This can mean anything from undesirable off-flavours or aromas right through to having to ditch a 20L down the drain, and that is heartbreaking. Actually it's worse than heartbreaking.

Many people are put off brewing beer at home from having tried samples in the past from friends that had a certain 'home-brew twang'. This is invariably caused by poor sanitisation.

Now, you can't sanitise something unless it is clean in the first place. There's no point using a sanitiser solution on a crusty, dirty brewing bucket, so your first step is to clean your kit. Then you sanitise it.

SO, STEP 1 – CLEAN; STEP 2 – SANITISE

The good news is that a couple of products – Star-San and Oxi-Clean – exist that make cleaning and sanitisation very quick and easy. You just need a process and, handily enough, I've written it for you.

You may already have Oxi-Clean in your laundry cupboard. It's the stuff used to get stains out of your dirty washing but it serves as a very powerful cleaning agent for home-brew equipment and will also remove any gunge or gunk that is caked on to your kit. When you use it, it is important to rinse the equipment a few times. Buy the unperfumed version if you can – most supermarkets have the big 1kg tubs on offer at some point.

Star-San is an instant sanitiser that was developed for the meat, food and beverage industries. It kills bugs on contact (well, within a minute or two) and, conveniently for us lazy brewers, you don't have to rinse it off afterwards. How cool is that? It's simple to use – you just mix up as per instructions on the side of the bottle and then give whatever needs sanitising a swirl around with it. Tip out the excess and you are good to go. Amazing stuff. The bottles last forever too.

CLEANLINESS IS NEXT TO GODLINESS

MY PROCESS

REMOVE EXCESS GUNK FROM YOUR KIT WITH A DAMP CLOTH, THEN WIPE OVER WITH KITCHEN ROLL.

ADD THREE SCOOPS OF OXI-CLEAN TO A FERMENTING BUCKET AND THEN HALF-FILL WITH LUKEWARM WATER.

MIX UNTIL ALL THE POWDER HAS DISSOLVED.

ADD ANY BREWING EQUIPMENT, PLACE LID ON FERMENTER AND SLOSH AROUND SO ALL INTERNAL SURFACES ARE CLEANED. REPEAT EVERY TEN MINUTES FOR THIRTY MINUTES.

DISCARD THE OXI-CLEAN AND RINSE BUCKET AND ALL EQUIPMENT THOROUGHLY ABOUT THREE TIMES UNTIL YOU CAN NO LONGER SMELL THE CLEANER. YOU CAN POUR THE USED OXI-CLEAN INTO ANOTHER FERMENTER OR KEG IF YOU LIKE. PROVIDING IT IS STILL FOAMING THEN IT'S STILL CLEANING.

MIX UP A SMALL BATCH OF STAR-SAN – 1L IS AMPLE.

POUR THIS MIX INTO YOUR FERMENTER, ADD YOUR BREW EQUIPMENT AND SLOSH AROUND SO EVERYTHING GETS COATED.

POUR OUT EXCESS. DON'T WORRY ABOUT RESIDUAL BUBBLES AND DO NOT RINSE. DON'T FEAR THE FOAM.

THAT'S IT. EASY.

I ALSO FILL A CHEAP SPRAY BOTTLE WITH A SOLUTION OF STAR-SAN. THAT WAY, I CAN SPRITZ ANY ODD BIT OF EQUIPMENT THROUGHOUT THE BREWDAY.

EVERYTHING THAT TOUCHES YOUR BEER AFTER THE BOIL MUST BE SANITISED.

3

USING A SMALLER POT

I started off brewing 19L batches of beer using a 13L kitchen stock pot. There are some good ways this can be easily done.

If you are brewing fairly low-gravity beers (5% or less) that do not require huge amounts of grain then you follow exactly the same steps as using a big pan, adding the entire amount of grain listed in the recipe at once at the start. At the end of the boil, you will have roughly 10L of highly concentrated wort that you can then dilute back down to 19L with tap water in your fermenter.

If you are brewing something a bit stronger, you are limited by how much grain and water you can fit in the pot at the same time. In this instance, stick with the maximum amount of grain you think you can fit (probably no more than 4kg) and brew as normal. Towards the end of the boil you then add sufficient dry or liquid malt extract to the boil to bring the beer up to the strength you are shooting for. Sugar can be used, but the sugar alone will only add alcohol and nothing by way of flavour, colour or mouth feel. It will leave you with a thin and dry beer.

Or you split all your ingredients in half and brew twice, combining the two batches together in the fermenter before you add the yeast. This works but makes for a very long brewday.

BASIC EQUIPMENT

IDEALLY YOU'LL HAVE MOST OF THE KIT YOU'LL NEED LYING AROUND THE KITCHEN OR SHED.

THE ONLY THINGS YOU MAY NEED TO PURCHASE ARE THE FERMENTING BUCKET AND OF COURSE THE BEER INGREDIENTS. IF YOU ARE LUCKY, YOU'LL HAVE A FRIENDLY LOCAL HOME-BREW SHOP THAT CAN FIND THIS KIT FOR YOU, OR ORDER ONLINE FROM THE MALT MILLER.

MUST-HAVE KIT

LARGE PAN

My first all-grain beers were made using a 13L stock pot that we used for making soups or boiling pasta. I made some fantastic beers using this pan. Even if you want to brew in 19L batches, there are several methods to get this volume out of a smaller pan. Any large stock pot can be used to brew beer. The important thing is it must be clean! I now use a 33L restaurant stock pot and that is a good size to make 19L-batches of beer.

DRAWSTRING MESH BAG

This is the key to the Brew-in-a-Bag method of making beer. You'll use this mesh bag twice during your brewday – once to steep your grains in the pot and then a second time to strain the beer before it goes into the fermenter. If you can afford it, buy two mesh bags as it saves a bit of hassle on your brew day. If you are handy with a sewing machine (or know someone who is) you can make your own mesh bag, but for £5–£10 you can buy one online that will do the trick.

THERMOMETER

I use a Thermapen instant-read digital thermometer because it is wonderful. Mine gets used lots for cooking meat too. Anything that gives an accurate reading within +/–2°C is fine. A jam thermometer will do at a pinch.

LARGE SPOON

Metal or plastic is fine. Wood is no good as it is hard to sanitise properly. Make sure it has a handle that is longer than the depth of your pan!

SCALES

Metal or plastic is fine. You'll need scales firstly to weigh out the right amount of grain and then secondly to measure out your hops. There should be no need for them to read over 5kg but being able to measure single grams accurately is important. I use Salter kitchen scales.

LARGE PLASTIC FERMENTER WITH LID AND AIRLOCK

This is the only bit of kit you'll probably need to buy. You can pick them up for virtually nothing at car boot sales but I would advise against this as they may have scratches that harbour bugs and make them hard to clean. Head for your local home-brew store or online. My fermentation buckets are around 27L. It needs to have a lid and ideally a little airlock.

SANITISER – THIS IS CRUCIAL!

As described on page 21, the two products I recommend are Oxi-Clean for getting rid of the muck and then Star-San for sanitising. These two products make life much simpler.

SIPHON TUBE

This is for transferring your beer from the vessels to the bottles.

NICE-TO-HAVE KIT

These are a few bits and pieces that make your brewing life a little easier. You certainly do not need them to make great beer so perhaps consider them after you've made a few batches and know your new hobby is a keeper.

HYDROMETER – this is used to measure how strong your beer is

BOTTLING WAND – makes bottling beer easier and less messy

BOTTLE CAPPER, CAPS & BOTTLES

THE
BELL
PULL
Stroud Brewery
Organic Ale
4%
£GRAND
IN YOUR HAND
WIN £1000
Stroud Brewery
Light Ale
3.6%
THE
LAST
DUEL
Stroud Brewery
Pale Ale
3.8%
BAUGHAN
TO RIDE
Stroud Brewery
Wheat Ale
4.2%
REDCOAT
Stroud Brewery
Ruby Mild
3.8%
RESOLUTION
Stroud Brewery
Single Hop Pale
4.2%
I.P.A.
HELLISH
4.2%

ČESKÁ REPUBLIKA
2015
Žatecko
FABRIC
VAUXHALL
SPARES U7

STEP-BY-STEP BREWDAY

FIRST THINGS FIRST. CLEAR AS MUCH SPACE IN YOUR KITCHEN AS YOU CAN.

GET ALL YOUR EQUIPMENT AND INGREDIENTS READY AND PUT SOME LOUD MUSIC ON.

AND POUR A . . .

BEER

MEASURE OUT YOUR GRAIN INTO TWO OR THREE BOWLS OR JUGS THAT YOU CAN EASILY LIFT WITH ONE HAND.

FILL YOUR BIG PAN ABOUT TWO-THIRDS FULL WITH WATER AND BRING TO AROUND 72°C. KILL THE HEAT AND STRETCH THE NECK OF YOUR MESH BAG AROUND THE TOP OF THE PAN.

IDEALLY YOUR MESH BAG WILL HAVE A DRAWSTRING THAT YOU CAN SECURE TIGHTLY.

ADD THE GRAIN INTO THE MESH BAG IN THE PAN IN A THIN STREAM, STIRRING CONSTANTLY TO AVOID PRODUCING DOUGH-BALLS. ONCE ALL THE GRAIN HAS BEEN ADDED, CONTINUE TO STIR UNTIL ALL LUMPS HAVE BEEN BROKEN UP. THE BIG LUMPS WILL FLOAT SO JUST SQUASH THEM AGAINST THE SIDE OF THE PAN.

CHECK THE TEMPERATURE OF THE RESULTING MIXTURE. WE USUALLY WANT OUR GRAIN TO STEEP AT BETWEEN 64 AND 66°C DEPENDING ON THE RECIPE. IF IT IS TOO HIGH, POUR IN SOME COLD WATER AND STIR IT AROUND.

IF YOU HAVE COME IN TOO LOW, THEN FIRE UP THE BURNERS UNDER THE PAN AND KEEP STIRRING. ONCE YOU HIT YOUR TARGET TEMPERATURE, CLOSE THE LID.

STIR WELL AND CHECK THE TEMPERATURE EVERY FIFTEEN MINUTES – IF IT HAS DROPPED OFF, FIRE UP THE BURNERS FOR A MINUTE OR TWO.

LEAVE THE GRAINS TO STEEP FOR SIXTY TO NINETY MINUTES. IN THE BREWING WORLD THIS IS KNOWN AS MASHING.

FIRE UP THE BURNERS AND RAISE THE TEMPERATURE TO 75°C, THEN KILL THE FLAMES AGAIN.

NEXT GRAB A SPARE BUCKET AND PLACE IT NEARBY.

CAREFULLY LIFT OUT YOUR BAG OF GRAIN. ALLOW AS MUCH FLUID AS POSSIBLE TO DRAIN OFF BEFORE DUMPING THE BAG IN YOUR BUCKET.

FIRE UP THE BURNERS UNDER YOUR BIG PAN TO START BRINGING YOUR HOT MALTY WATER (KNOWN AS WORT) UP TO BOILING POINT.

I MEASURE THE VARIOUS HOP ADDITIONS INTO SEPARATE BOWLS, ALL LINED UP IN ORDER WITH EACH TO BE ADDED AT A DIFFERENT TIME DURING THE BOIL.

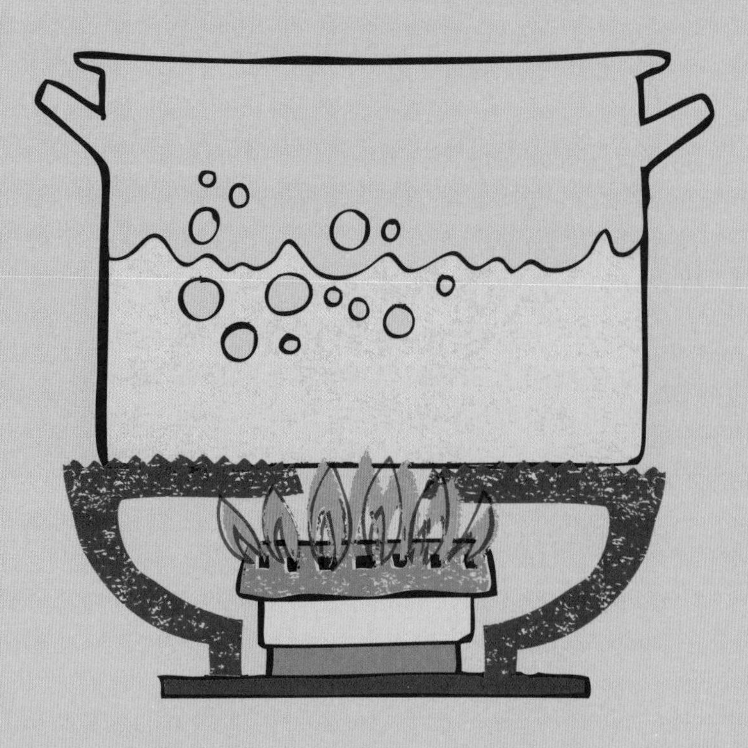

WHILE THE LIQUID IS HEATING UP IT IS CRUCIAL YOU STAY CLOSE BY AS IT HAS A HABIT OF SUDDENLY BOILING UP AND COVERING YOUR KITCHEN IN A HOT STICKY MESS. NOT COOL.

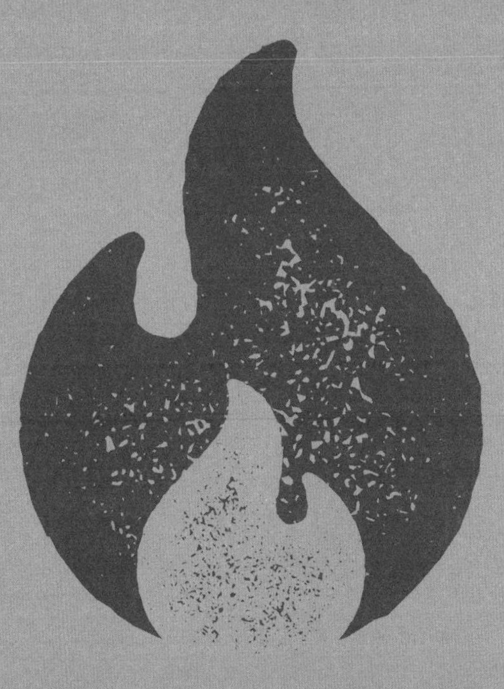

IF IT LOOKS LIKE YOUR PAN IS ABOUT TO BOIL OVER, TURN OFF THE HEAT, STIR RAPIDLY AND BLOW ON THE SURFACE. YOU CAN ALSO HAVE A SPRAY BOTTLE TO HAND AND SPRITZ THE SURFACE IF IT LOOKS A LITTLE LIVELY. YOUR BEST BET IS NOT TO FILL YOUR PAN UP RIGHT TO THE TOP IN THE FIRST PLACE – GIVE YOURSELF A FEW INCHES' BUFFER JUST IN CASE.

ONCE IT IS AT BOILING POINT AND IT HAS CALMED DOWN A LITTLE BIT, YOU CAN ADD YOUR FIRST HOPS AND START YOUR COUNTDOWN TIMER. CONTINUE TO ADD THE REST OF THE HOPS THROUGHOUT THE BOIL ACCORDING TO THE TIMINGS GIVEN IN YOUR CHOSEN RECIPE. EARLIER HOPS GIVE BEER THE BITTER FLAVOUR, HOPS ADDED IN THE MIDDLE ADD TO THE FLAVOUR, WHEREAS HOPS ADDED AT THE END WILL GIVE THE AROMA.

ONCE YOU HAVE REACHED THE END OF YOUR BOIL TIME, YOU NEED TO COOL YOUR BEER AS QUICKLY AS YOU CAN.

THE LONGER IT SITS AROUND HOT AND UNPROTECTED, THE MORE LIKELY IT IS THAT NASTIES WILL TAKE HOLD AND RUIN YOUR BEER.

ALSO, THE QUICKER YOU CAN COOL IT, THE MORE OF THE DELICATE AND DELICIOUS HOP AROMAS WILL BE PRESERVED.

MY CHOSEN METHOD IS TO FILL THE KITCHEN SINK (OR YOUR BATH IF CONVENIENT) WITH COLD WATER THEN IMMERSE THE PAN AS DEEP AS YOU CAN WITHOUT YOUR SINK OVERFLOWING.

STIR REGULARLY, MAKING SURE YOU DO NOT REMOVE THE SPOON AT ANY TIME – YOU DON'T WANT IT TO PICK UP BUGS. ONCE THE SURROUNDING WATER STARTS TO HEAT UP I PULL THE PLUG FROM THE SINK AND CAREFULLY LET THE COLD-WATER TAP RUN. I AIM TO GET MY BEER DOWN TO APPROXIMATELY 25°C BEFORE IT'S READY TO MOVE.

REMEMBER THAT ANYTHING THAT COMES INTO CONTACT WITH YOUR BEER FROM THE MOMENT YOU TURN THE BURNERS OFF MUST BE SANITISED.

YOU CAN ALSO HELP ELIMINATE THE CHANCE OF NASTIES BY GIVING YOUR SINK AND SURROUNDING AREA A GOOD CLEAN WITH ANTIBACTERIAL SPRAY BEFORE YOU START.

NEXT, SECURE A CLEAN AND FRESHLY SANITISED MESH BAG AROUND THE COLLAR OF A CLEAN AND SANITISED FERMENTING VESSEL. I BREW 20L BATCHES SO I CAN LIFT AND POUR MY PAN OF WATER BUT YOU MAY NEED TO USE A SIPHON TUBE IF BREWING BIGGER BATCHES.

AT THIS POINT IT IS BENEFICIAL TO GET OXYGEN INTO YOUR BEER SO I POUR FROM A HEIGHT OF AROUND 2 FEET TO CREATE A GOOD SPLASH IN THE BUCKET. THE MESH BAG COLLECTS ALL THE HOPS AND OTHER DEBRIS.

LIFT OUT THE MESH BAG AND LET THE LIQUID DRAIN OFF. I USUALLY GIVE IT A SQUEEZE (WITH CLEAN HANDS) AS THE LEAF HOPS ARE LIKE LITTLE SPONGES AND TRY TO STEAL YOUR BEER.

REGARDLESS OF THE SIZE OF POT YOU HAVE USED SO FAR, NOW IS THE TIME TO CORRECT THE AMOUNT OF BEER IN YOUR FERMENTING VESSEL. YOU ARE AIMING FOR 19L AT THE END OF FERMENTATION AND UNLESS YOU ARE USING A HUGE POT, IT IS LIKELY YOU'VE NOT GOT ENOUGH BEER AT THE MOMENT. ALL YOU DO NOW IS TOP UP YOUR FERMENTER TO AROUND THE 20L MARK WITH COLD TAP WATER. READ THE SECTION ON BREWING IN SMALLER POTS (SEE PAGE 24) FOR MORE INFO.

THE RESULTING BEER IS NOW READY TO HAVE THE YEAST ADDED BEFORE THE LID IS SEALED AND AN AIRLOCK FITTED.

STASH THE BEER SOMEWHERE WITH A CONSISTENT AMBIENT TEMPERATURE OF AROUND 18 TO 22°C. TOO HOT AND YOUR BEER WILL DEVELOP OFF-FLAVOURS, TOO COOL AND YOUR YEAST WILL FALL ASLEEP.

LAST JOB (APART FROM THE CLEAN-UP) IS TO KICK BACK, RELAX AND POUR YOURSELF ANOTHER BEER. YOU DESERVE IT. FIND A FRIEND TO PAT YOU ON THE BACK TOO.

MOST REGULAR-STRENGTH BEERS WILL BE READY TO DRINK IN THREE TO FOUR WEEKS. IDEALLY YOU'LL HAVE THE PATIENCE TO WAIT SIX TO EIGHT WEEKS OR EVEN UP TO TWELVE WEEKS, AS THEY WILL IMPROVE WITH TIME. BUT NOBODY HAS THAT WILLPOWER. I CERTAINLY DO NOT.

BREWDAY SUMMARY

1. MEASURE OUT YOUR GRAIN ACCORDING TO YOUR RECIPE.

2. ADD 27L (OR AS CLOSE AS YOU CAN) OR SO OF WATER TO YOUR BREWPOT, FIRE UP THE BURNERS AND BRING IT TO 72°C, THEN TURN OFF THE HEAT.

3. FASTEN YOUR MESH BAG TO THE RIM OF THE PAN AND THEN SLOWLY TIP IN YOUR GRAIN, STIRRING GENTLY TO AVOID DOUGH-BALLS.

4. MAINTAIN A TEMPERATURE OF 66°C FOR AN HOUR. IF THE TEMPERATURE DROPS THEN GIVE THE PAN A QUICK BLAST OF HEAT.

5. NOW FIRE THE BURNERS BACK UP AND BRING THE TEMPERATURE OF THE LIQUID UP TO 75°C.

6. REMOVE THE MESH BAG AND DUMP IT IN A BUCKET.

7. FIRE THE BURNERS UP AND BRING THE LIQUID TO A RAPID BOIL.

8. ADD YOUR HOPS AT THE TIMES GIVEN IN YOUR RECIPE.

9. COOL YOUR BEER TO 25°C AND TRANSFER INTO A CLEAN AND SANITISED FERMENTING BUCKET.

10. ADD YEAST.

11. SEAL THE LID, ADD THE AIRLOCK AND STASH SOMEWHERE TO FERMENT FOR TWO TO THREE WEEKS.

FOLLOWING AND CREATING RECIPES

UNDERSTANDING BREWING RECIPES

I PROMISED RIGHT AT THE START OF THIS BOOK NOT TO USE COMPLEX JARGON, OR TOO MUCH BREWING TERMINOLOGY. LET'S ASSUME YOU REALLY GET INTO BREWING AND START TRAWLING THE WONDERFUL WORLDWIDE WEB FOR OTHER NEW AND EXCITING BEER RECIPES. THEN IT IS IMPORTANT YOU CAN GET YOUR HEAD ROUND HOW THEY WORK. SO THIS SECTION IS ABOUT HELPING YOU TO UNDERSTAND OTHER PEOPLE'S BEER RECIPES.

MOST BEER RECIPES FOLLOW A SIMILAR TEMPLATE THAT CAN BE CONFUSING TO START WITH, BUT YOU'LL GET THE HANG OF THEM PRETTY QUICKLY AND BE ABLE TO START PUTTING YOUR OWN RECIPES TOGETHER TOO.

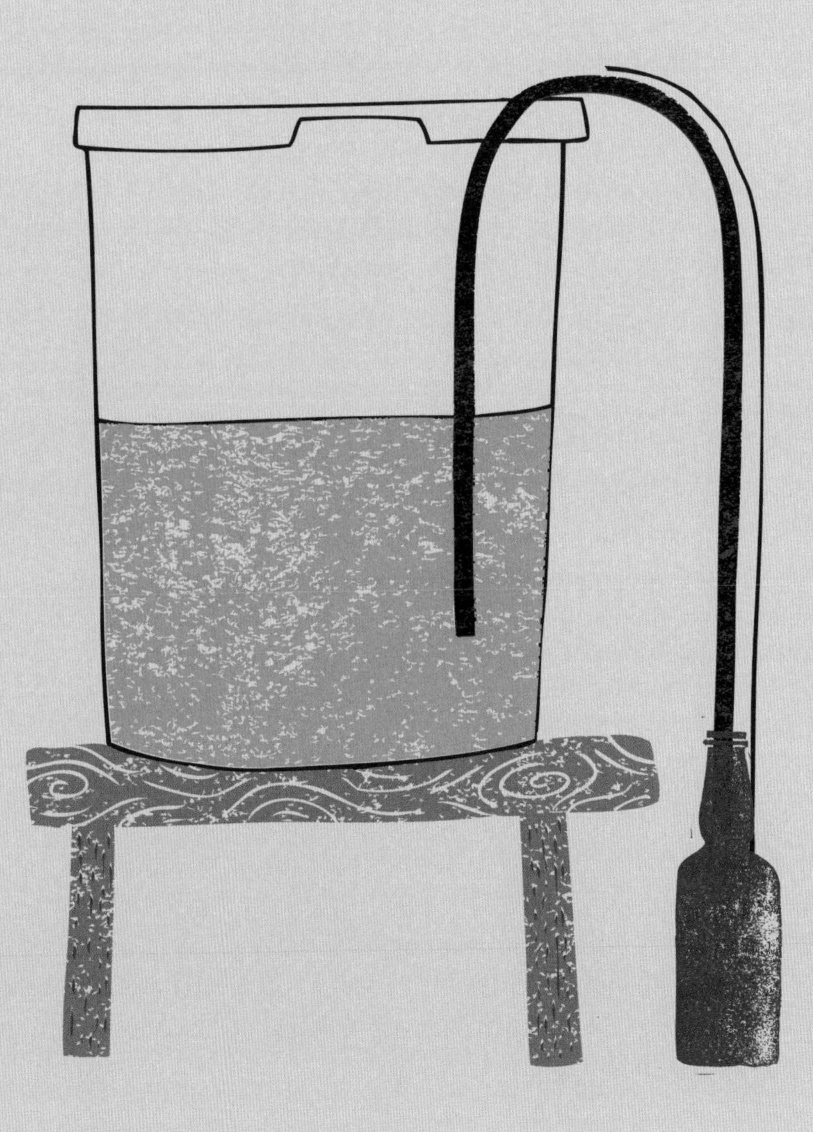

THE KEY INFORMATION IS CONTAINED IN A NUMBER OF MEASURED UNITS AS FOLLOWS:

GRAVITY READINGS

OG – THE ORIGINAL GRAVITY OF THE BEER

This tells us how much fermentable sugar is present in our beer before we add the yeast. The higher the number, the more sugar is present for the yeast to eat and therefore the stronger the beer is likely to be.

FG – THE FINAL GRAVITY OF THE BEER

This is how much sugar is left in the beer once fermentation is complete. The lower the number, the more sugar has been consumed and turned to alcohol and therefore the stronger and drier the beer. A higher FG usually means the beer has more 'mouth feel' and is a little sweeter, often maltier. Too low and the beer can feel thick and a bit watery.

IF YOU WANT TO CHECK THE GRAVITY OF YOUR BEERS, USE A HYDROMETER THAT YOU CAN BUY CHEAPLY IN YOUR LOCAL HOME-BREW STORE. REMEMBER TO SANITISE BEFORE YOU USE IT!

DIFFERENT BEER STYLES DICTATE DIFFERENT ORIGINAL AND FINAL GRAVITIES. MY AMERICAN PALE ALE STARTS WITH AN OG OF 1.050 AND FINISHES AROUND 1.010.

ONCE WE KNOW THE STARTING AND FINAL GRAVITY OF THE BEER, WE KNOW HOW MUCH SUGAR HAS BEEN CONVERTED TO ALCOHOL AND THEREFORE THE STRENGTH OF THE BEER. OUR PALE ALE IS GOING TO FINISH AT AROUND 5.2% ABV (ALCOHOL BY VOLUME).

MASH SCHEDULE

This is a fancy name for the grain you will use in your recipe. Typically you'll be using a 'base' malt that will account for most of the grain and then adding a number of 'speciality' malts that will impart different flavours, colours and aromas to your beer. Stout, for example, will still use 90% 'pale malt' but with the addition of 10% dark roasted malt to give it its trademark colour, smell and roasted flavour.

BOIL

This simply determines how long you boil the beer for. As you know, hops will be added at various times throughout the boiling process and this is usually denoted in the recipe by giving the amount of hops to add (in grams) and then how long before the end of the boil to add them. For example, 40g Cascade at 45 minutes means you add 40g of Cascade hops forty-five minutes before the end of the boil. 20g Cascade at 5 minutes would mean you add the hops five minutes before the end of the boil.

IBU/EBU – BITTERNESS UNITS

These are the same thing and are a measure of the bitterness in beer given by the hops. Our pale ale is aiming for an IBU of 36. Some huge American IPAs can be up in the 70s and 80s.

SRM/EBC

These both measure the colour or darkness of beer. The higher the number, the darker the beer. Essentially they work the same way, with SRM being the US measurement and EBC the UK. As long as you stick with one or the other for your own records you'll be able to compare and measure your beers. I never bother recording mine. Frankly I don't really care what colour my beer is as long as it looks and tastes good.

SMASH BEERS

SMASH BEERS ARE GREAT. THEY ARE A FUN WAY TO REALLY GET TO KNOW YOUR MALTS AND YOUR HOPS BUT ALSO IT SOUNDS COOL. SMASH. SOUNDS A BIT ROCK AND ROLL I THINK.

THE CONCEPT IS SIMPLE – SINGLE MALT AND SINGLE HOP.

SO BASICALLY YOU CHOOSE A BASE MALT YOU LIKE THE SOUND OF AND THEN YOU FIND YOURSELF A NICE ALL-ROUND HOP. AND THAT'S ALL YOU USE. IT WON'T GIVE YOU THE MOST INTERESTING COMPLEX BEER IN THE WORLD, BUT IT WILL HELP YOU REALLY DRILL DOWN INTO THE SPECIFIC CHARACTERISTICS OF THOSE INGREDIENTS SO YOU CAN START BUILDING UP YOUR KNOWLEDGE BANK OF FLAVOURS.

A POPULAR SMASH COMBO TO KICK OFF WITH WOULD BE MARIS OTTER MALT AND CASCADE HOPS.

NOW RIFF ON IT. MAYBE SWAP OUT THE CASCADE HOP FOR A MORE PUNCHY NEW ZEALAND HOP SUCH AS NELSON SAUVIN AND NOTICE THE DIFFERENCE IN THE BEER. OR SWAP OUT THE YEAST FOR A DIFFERENT YEAST.

SMALL BATCHES OF SMASH BEERS ARE REALLY QUICK AND EASY. GET ON IT!

HEFEWEIZEN
PECKHAM PALE
CIDER.

HOME-BREW RECIPES

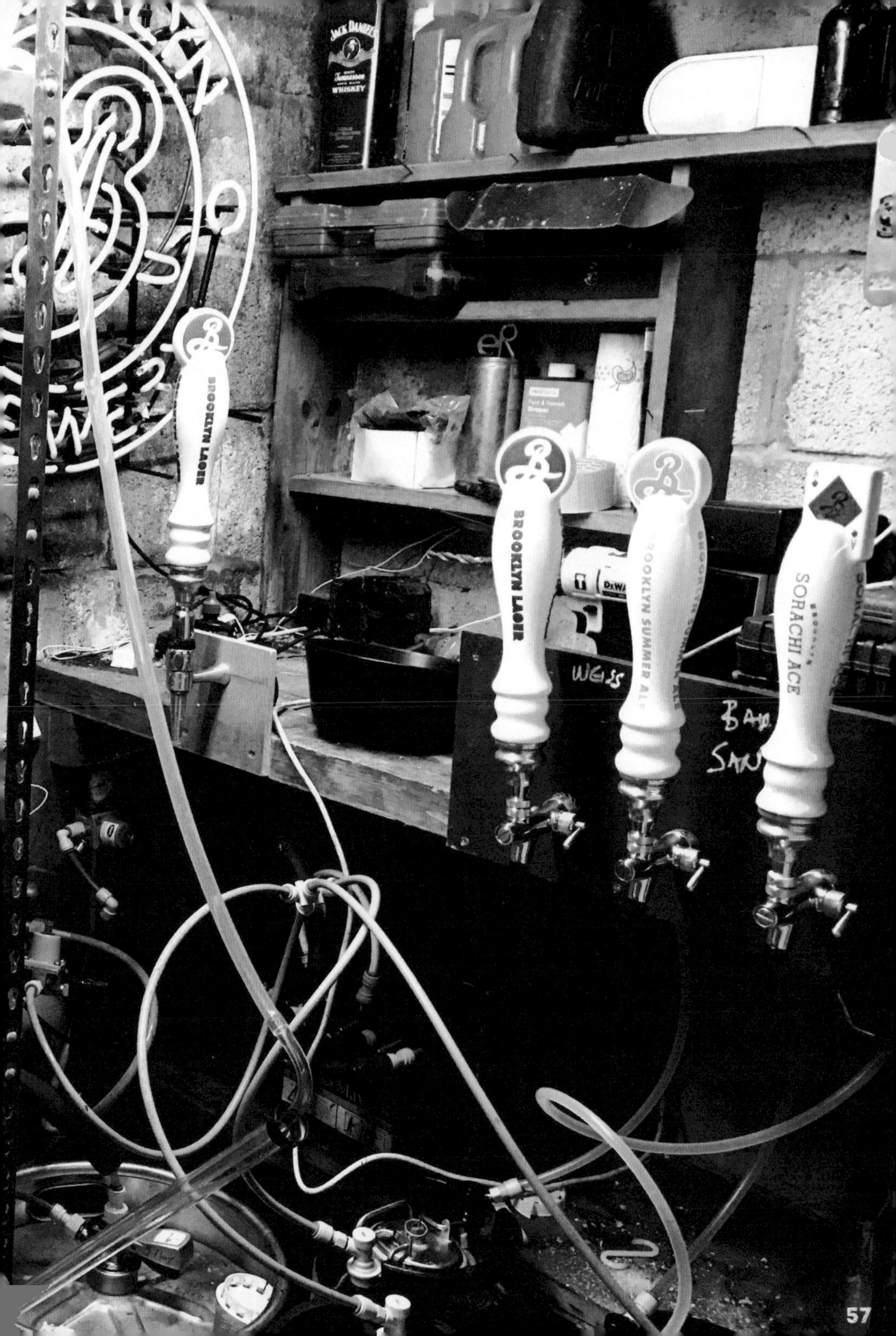
JACK DANIEL'S
WHISKEY
BROOKLYN LAGER
BROOKLYN LAGER
BROOKLYN SUMMER ALE
SORACHI ACE
WEISS

WE'VE ALREADY RUN THROUGH HOW TO FOLLOW A HOME-BREW RECIPE. THE KEY IS TO IGNORE THE JARGON, FIGURE OUT HOW MUCH GRAIN YOU ARE STEEPING AND THEN WHAT HOPS YOU WILL BE ADDING DURING THE BOIL.

THE PROCESS FOR MOST RECIPES REMAINS VIRTUALLY THE SAME THROUGHOUT.

You may wish to use malt extract instead of brewing 'all-grain' and of course that's fine. You can swap out the 'base' malt in these recipes at an approximate ratio of:

1kg grain = 0.75kg liquid malt extract
1kg grain = 0.6kg dry malt extract

So if we plan to brew with dry malt extract and a recipe calls for 5kg of pale ale grain as its base, then we multiply by 0.6 and substitute around 3kg of extract.

Flame-out hops should be added once you have taken the pan off the hob and started to cool it. They'll get strained off when you transfer your beer to the bucket half an hour later. We're adding them at this stage so all the volatile aromas do not get boiled off.

Remember, if a recipe calls for dry hopping, then we're adding those hops to the fermenting bucket around one week after you've made your beer (see page 14). Aim to dry hop for around one further week unless the recipe tells you otherwise. Set yourself a reminder!

The following recipes are intended to give you a few first brews to play around with. There are tons of superb home-brew recipe publications out there and I've listed my favourites at the end of this book for when you are ready to move on.

RED BRICK RYE

TARGETS:
ORIGINAL GRAVITY: 1.048
FINAL GRAVITY: 1.011
ABV: 5.0%

BOIL TIME: 75 MIN
BATCH SIZE: 19L

INGREDIENTS

GRAIN
2.7kg – Pale Malt
500g – Carared
700g – Rye malt
60g – Roasted barley

HOPS
16g – Columbus – Boil 75 min
10g – Chinook – Boil 10 min
5g – Cascade – 5 min
7g – Chinook – Flame-out
15g – Cascade – Flame-out

YEAST
1 pack Safale US–05

AFTER FERMENTATION COMPLETE DRY HOP WITH: 100g CHINOOK IN THE FERMENTING BUCKET AND LEAVE FOR A FURTHER FOUR DAYS.

RECIPE KINDLY GIVEN BY BRICK BREWERY.

HOUSE PALE ALE

TARGETS:
ORIGINAL GRAVITY: 1.049
FINAL GRAVITY: 1.010
ABV: 5.2%
COLOUR: 13.6 EBC
BITTERNESS: 47.6 IBU

BOIL TIME: 60 MIN
BATCH SIZE: 19L

INGREDIENTS

GRAIN

3.5kg – Pale Malt, Maris Otter
500g – Caramel Malt

HOPS

20g – Northern Brewer Hops – Boil 60 min
20g – Cascade Hops – Boil 45 min
20g – Cascade Hops – Boil 10 min
50g – Cascade Hops – Flame-out

YEAST

1 pack Safale English Ale #S–04 Yeast

AMERICAN PALE ALE IS A GREAT BEER TO HAVE HANDY AT ALL TIMES IN MY VIEW. EVERYONE LIKES IT. IT'S REFRESHING AND CLEAN, AROMATIC AND FLAVOURSOME BUT WITHOUT BEING EXTREME IN ANY WAY. IT'S BEAUTIFULLY BALANCED AND EASY TO MAKE. IT'S ALSO NOT TOO STRONG, SO IT IS A GOOD ONE TO GET STUCK INTO AT THE WEEKEND.

HERE'S A SIMPLE RECIPE THAT GETS REAL CHARACTER FROM THE ADDITION OF A SMALL AMOUNT OF CARAMEL MALT GIVING A REALLY BEAUTIFUL COLOUR AND DELICIOUS MALTY BACKBONE. IT ALSO SEES THE ADDITION OF HOPS JUST AS YOU TURN OFF THE HEAT AT THE VERY END OF BOILING THE BEER. ADDING HOPS THIS LATE DELIVERS A HUGE WHACK OF HOP AROMA BUT VIRTUALLY NOTHING BY WAY OF BITTERNESS.

ACE OF SPADES
LONDON PORTER

TARGETS:
ORIGINAL GRAVITY: 1.050
FINAL GRAVITY: 1.014
ABV: 4.7%

BOIL TIME: 90 MIN
BATCH SIZE: 19L

INGREDIENTS
GRAIN
MASH
3kg – Maris Otter
300g – Chocolate Malt
150g – Roasted Barley
250g – Crystal 150
120g – Torrified Wheat

HOPS
12g – Centennial – Boil 90 min
18g – Williamette – Boil 30 min
30g – Williamette – Boil 5 min

YEAST
1 Pack Irish Ale Yeast

BAD SANTA

HERE'S A BEER I BREWED FOR THE FIRST TIME ABOUT FIVE YEARS AGO TO OFFER HOUSE GUESTS BUT ALSO TO BOTTLE AND GIVE AWAY AS GIFTS TO FRIENDS AND FAMILY.

MANY CHRISTMAS BEERS ARE OVERPOWERING AND AKIN TO DRINKING A GLASS OF CHRISTMAS PUDDING. AWESOME FOR THE FIRST MOUTHFUL OR TWO BUT THEY THEN START TO GET A BIT HEAVY, THICK AND SWEET. I WANTED TO BREW SOMETHING MORE BALANCED THAT TASTED FESTIVE BUT THAT YOU ENJOYED TO THE BOTTOM OF THE GLASS AND THAT MADE YOU REACH FOR ANOTHER. A GOOD BEER THAT CELEBRATED CHRISTMAS.

THIS IS THE MOST COMPLEX BEER I'VE LISTED HERE, BUT IT'S REALLY WORTH THE EFFORT AND THE INVESTMENT IN EXTRA INGREDIENTS. I RECOMMEND BREWING THIS AT THE START OF OCTOBER TO GIVE IT TIME TO CONDITION PROPERLY AND REALLY LET THE FLAVOURS DEVELOP.

TARGETS:
ORIGINAL GRAVITY: 1.064
FINAL GRAVITY: 1.013
ABV: 6.7%
COLOUR: 26.1 EBC
BITTERNESS: 23.4 IBU

BOIL TIME: 60 MIN
BATCH SIZE: 19L

INGREDIENTS

GRAIN

4kg – Pale Malt, Maris Otter
370g – Caramel/Crystal Malt
370g – Caramunich Malt
370g – Munich Malt
20g – Black (Patent) Malt
350g – maple syrup

HOPS/OTHER FLAVOURS

27.5g Northern Brewer Hops – Boil 60 min
1 vanilla bean – Boil 20 min
3 cinnamon sticks – Boil 20 min

YEAST

1 pack Irish Ale Yeast White Labs #WLP004 yeast

FINALLY ADD 250ml OF CHERRY EXTRACT AFTER TWO WEEKS

AMERICAN IPA

THE HISTORY BEHIND THE NAME INDIA PALE ALE IS PRETTY WELL KNOWN NOW.

IT HARKS BACK TO THE DAYS OF THE BRITISH EMPIRE, WHEN EXPATS WERE POSTED IN INDIA FOR LONG PERIODS. AT THE TIME INGREDIENTS FOR BEER WERE HARD TO FIND IN INDIA, AND BESIDES, THE AMBIENT TEMPERATURE WAS FAR TOO HOT TO BREW THEIR FAVOURITE TIPPLE – BEER!

WORD WAS SENT BACK TO THE THEN LEGENDARY BREWERS OF BURTON-UPON-TRENT TO MAKE BEER AND SHIP IT OVER TO INDIA.

The story goes that when the brewers were planning the beer, they carefully considered the three-month sea journey the product had to endure. Most beer would end up spoiled in this time so they designed the beer to be as resilient to bacteria as possible. Hops themselves have high antiseptic properties so they packed them into the brew. This combined with a much higher alcohol level than beers being brewed at the time gave it the best chance possible of surviving the trip in a drinkable condition. Happily, what they discovered was three months of conditioning mellowed out the initial harshness of the hops, leaving a deliciously hoppy, big-flavoured and refreshing beer perfectly suited to the Indian expat lifestyle and the strong flavours of Indian cuisine.

MASTERING YOUR OWN IPA IS A RITE OF PASSAGE FOR ANY HOME BREWER.

THE TRICK IS TO ALLOW IT TO CONDITION FOR A LITTLE LONGER THAN YOU WOULD WITH OTHER BEERS. THIS IN ITSELF IS DIFFICULT IF YOU ARE AS IMPATIENT AS ME, SO I SUGGEST YOU HAVE PLENTY OF BEER AROUND TO DRINK WHILE THIS ONE CONDITIONS.

TARGETS:
ORIGINAL GRAVITY: 1.066
FINAL GRAVITY: 1.010
ABV: 7.3%
COLOUR: 21.9 EBC
BITTERNESS: 50.5 IBU

BOIL TIME: 60 MIN
BATCH SIZE: 19L

INGREDIENTS

GRAIN

5kg – Pale Malt, Maris Otter
300g – Caramel/Crystal Malt

HOPS

40g – Centennial Hops – Boil 60 min
30g – Centennial Hops – Boil 10 min
30g – Amarillo Gold Hops – Boil 0 min
30g – Simcoe Hops – Flame-out

YEAST

1 pack Safale American #US–05 Yeast

I HAD THE PLEASURE OF VISITING THE ERDINGER BREWERY NEAR MUNICH SOME TIME AGO. CLEARLY MUCH SAMPLING TOOK PLACE AND I REALLY FELL IN LOVE WITH THIS STYLE OF BEER.

IT'S SIMPLE TO MAKE, WITH ONLY ONE HOP ADDITION. IT FERMENTS AND CONDITIONS QUICKLY SO IT IS A GOOD BEER FOR IMPATIENT BREWERS.

THE MAGIC COMES FROM USING A PROPER BAVARIAN WHEAT BEER YEAST – THE SOURCE OF ALL THE BEAUTIFULLY BALANCED BANANA AND CLOVE AROMAS. THE YEAST COSTS A LITTLE MORE THAN REGULAR DRIED YEAST BUT IT MAKES A BIG DIFFERENCE HERE.

AS WITH ANY BEER, USE THE FRESHEST AND BEST INGREDIENTS YOU CAN FIND. THIS BEER IN PARTICULAR WILL HIGHLIGHT ANY INFERIOR INGREDIENTS.

TARGETS:

ORIGINAL GRAVITY: 1.050
FINAL GRAVITY: 1.009
ABV: 5.9%
COLOUR: 7.4 EBC
BITTERNESS: 11.5 IBU

BOIL TIME: 60 MIN
BATCH SIZE: 19L

INGREDIENTS

GRAIN

2kg – Wheat Malt, Ger
1.8kg – Pale Malt, Maris Otter
130g – Munich Malt

HOPS

25g – Hallertauer Hersbrucker – Boil 60 min

YEAST

1 pack Bavarian Wheat Yeast (Wyeast Labs #3056)

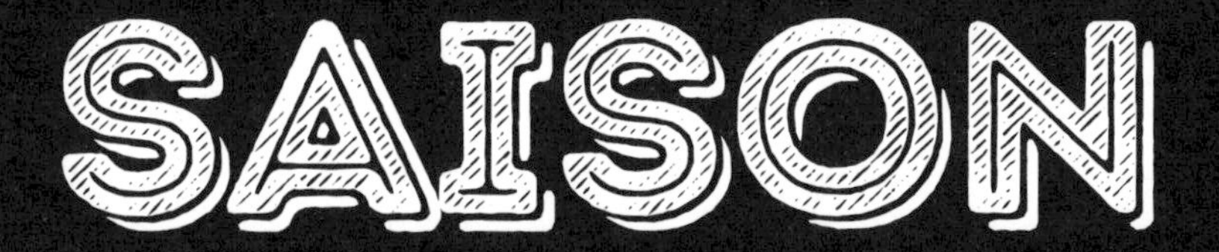

SAISON

THE 'SAISON' BEER STYLE ORIGINATED FROM BEERS BREWED DURING THE COOLER AND LESS ACTIVE MONTHS IN FARMHOUSES IN WALLONIA, THE FRENCH-SPEAKING REGION OF BELGIUM, AND THEN STORED FOR DRINKING BY THE FARM WORKERS DURING THE SUMMER MONTHS. EACH WORKER WAS ENTITLED TO A STAGGERING 5L PER DAY!! WHERE DO I SIGN UP??

THERE'S NO SINGLE DEFINITIVE 'SAISON'; INSTEAD, THE TERM SEEMS TO GROUP TOGETHER A COLLECTION OF AROMATIC AND REFRESHING SUMMER ALES. THIS LEAVES RECIPE DEVELOPMENT QUITE OPEN TO CREATIVITY IN TERMS OF CHOICE OF HOPS BUT ALSO IN CHOICE OF HERB AND SPICE ADDITIONS. TYPICALLY IT IS BREWED WITH A PILSNER MALT BUT YOU CAN USE ANY PALE MALT. IT IS HIGHLY CARBONATED AND DRY WITH AROMAS OF ORANGES, LEMONS AND FLORAL HOPS. A HINT OF PEPPERY SPICE IS PRESENT TOO, USUALLY COMING FROM THE YEAST.

WHAT COULD BE BETTER SUITED TO QUENCH THE THIRST ON A HOT SUMMER AFTERNOON THAN A BEER DESIGNED TO REFRESH A TIRED, HOT FARM WORKER?

TARGETS:
ORIGINAL GRAVITY: 1.047
FINAL GRAVITY: 1.008
ABV: 5.2%
COLOUR: 3.8 EBC
BITTERNESS: 41.9 IBU

BOIL TIME: 60 MIN
BATCH SIZE: 19L

INGREDIENTS

GRAIN

4kg – Pilsner Malt

HOPS

20g – Centennial Hops – Boil 60 min
40g – Saaz Hops – Boil 30 min
50g – Cascade Hops – Boil 5 min
20g – Whole coriander seeds – Boil 5 min
2g – Whole black peppercorns – Boil 5 min
1 zest of one orange (no pith) – Boil 5 min

YEAST

1 pack Safale American #US–05 Yeast or a Saison yeast

NE IPA

THE NEW KID ON THE BLOCK, NEBRASKA IPA (SOMETIMES MISTAKENLY REFERRED TO AS NEW ENGLAND IPA) IS IN MY VIEW A BIT OF A DIAL BACK FROM THE HUGE AMERICAN IPA HOP-BOMBS THAT BECAME POPULAR. HOME BREWERS AND PROFESSIONAL BREWERS ALIKE HAVE BEEN WAGING A HOP WAR ON WHO CAN MAKE A BEER WITH THE MOST HOPS CRAMMED IN. THE ISSUE IS YOU EVENTUALLY GET TO THE POINT WHERE THE BEER BECOMES SO BITTER AND RESINOUS THAT IT IS UNPLEASANT TO DRINK.

THAT'S WHERE THE NE IPA COMES IN. SURE, IT IS A HOP-BOMB, BUT IT DOESN'T HAVE THE EXTREME BITTERNESS USUALLY ASSOCIATED WITH IPAS. INSTEAD, IT IS ALL ABOUT THE HOP AROMA. TO ACHIEVE THIS, ALL THE HOPS ARE PUT IN EITHER RIGHT AT THE END OF THE BOIL OR AFTERWARDS BY DRY HOPPING.

A GOOD NE IPA SHOULD BE BALANCED, SOFT, JUICY AND SMOOTH. TONS OF HOPS, BUT NOT TOO BITTER. MOST IMPORTANTLY IT SHOULD HAVE A HAZE – ITS TELL-TALE SIGN.

TARGETS:
ORIGINAL GRAVITY: 1.072
FINAL GRAVITY: 1.014
ABV: 7.4%
COLOUR: 13.4 EBC
BITTERNESS: 30 IBU

BOIL TIME: 60 MIN
BATCH SIZE: 19L

INGREDIENTS

GRAIN

5kg – Pale ale malt
300g – Caramel malt
300g – Wheat malt
300g – Rolled oats

HOPS

30g – Citra – Boil 10 min
30g – Galaxy – Boil 10 min
40g – Citra – Flame Out
40g – Galaxy – Flame-out
40g – Mosaic – Flame-out
40g – Citra – Dry hop for 7 days
40g – Galaxy – Dry hop for 7 days
40g – Mosaic – Dry hop for 7 days

YEAST

1 pack London Ale Yeast

OATMEAL STOUT

THE UNLIKELY ADDITION OF OATS TO THIS BEER IS GOING TO MAKE YOUR STOUT SMOOTH, CREAMY AND WITHOUT AFFECTING THE FLAVOUR TOO MUCH.

IT SOUNDS LIKE AN AMBITIOUS BEER TO BREW BUT IT ISN'T. IT'S A TASTY ONE THOUGH.

TARGETS:
ORIGINAL GRAVITY: 1.062
FINAL GRAVITY: 1.016
ABV: 6.0%
COLOUR: 57.4 EBC
BITTERNESS: 33.7 IBU

BOIL TIME: 60 MIN
BATCH SIZE: 19L

INGREDIENTS

GRAIN

4kg – Pale Ale Malt
400g – Crystal Malt
400g – Chocolate Malt
450g – Rolled oats

HOPS

60g East Kent Goldings – Boil 60 min

YEAST

1 pack English Ale Yeast

OTHER

1 Irish Moss tablet

CALIFORNIA ★ COMMON ★

THE MOST FAMOUS EXAMPLE OF A CAL COMMON IS ANCHOR STEAM FROM SAN FRANCISCO. I CAN'T THINK OF ANOTHER BEER SO UNIVERSALLY LIKED. YOU CAN PASS A BOTTLE TO PRETTY MUCH ANYONE FROM DIE-HARD STELLA FANS TO COMPLETE CRAFT-BEER GEEKS AND THEY'LL ENJOY IT. IT HITS THAT PERFECT BALANCE BETWEEN BEING REFRESHING LIKE A PILSNER AND WHOLESOME LIKE AN ALE.

CAL COMMON OR STEAM BEER IS EASY TO MAKE AND PRODUCES A GREAT-TASTING, UNIQUE AND INTERESTING BEER FROM SIMPLE INGREDIENTS. IT'S A GOOD NEXT-STEP BEER TO MAKE ONCE YOU HAVE MASTERED YOUR PALE ALES. THE METHOD IS TO BREW AN ALE USING A LAGER YEAST BUT AT ALE FERMENTATION TEMPERATURES. LAGERS ARE FERMENTED MUCH COLDER THAN ALES. THIS HIGHER FERMENTATION TEMPERATURE PUTS THE YEAST INTO OVERDRIVE.

THERE ARE A NUMBER OF THEORIES AS TO WHERE THE NAME 'STEAM BEER' COMES FROM. ACCORDING TO THE ANCHOR BREWERY WEBSITE THE NAME COMES FROM THE ORIGINAL PRACTICE OF FERMENTING THE BEER, WITHOUT REFRIGERATION EQUIPMENT USUALLY NEEDED FOR MAKING LAGER, ON SAN FRANCISCO'S ROOFTOPS IN A COOL CLIMATE. IN LIEU OF ICE, THE FOGGY NIGHT AIR NATURALLY COOLED THE FERMENTING BEER, CREATING STEAM OFF THE WARM OPEN PANS.

OR IT MAY BE THAT THE CARBON DIOXIDE PRESSURE PRODUCED BY THE PROCESS WAS VERY HIGH, AND SO IT WAS NECESSARY TO LET OFF 'STEAM' BEFORE ATTEMPTING TO DISPENSE THE BEER. WHO KNOWS?

TARGETS:
ORIGINAL GRAVITY: 1.046
FINAL GRAVITY: 1.010
ABV: 4.8%
COLOUR: 6.6 EBC
BITTERNESS: 31.2 IBU

BOIL TIME: 60 MIN
BATCH SIZE: 19L

INGREDIENTS

GRAIN

4kg – Pilsner Malt
500g – Munich Malt

HOPS

40g – Hallertauer Mittelfrueh – Boil 60 min
15g – Hallertauer Mittelfrueh – Boil 10 min
15g – Hallertauer Mittelfrueh – Flame-out

YEAST

1 pack Wyeast 2112 California Lager

OTHER

1 Irish Moss tablet

ONCE FERMENTED AND KEGGED, LET THE BEER CONDITION FOR ABOUT TWO WEEKS BEFORE YOU START DRINKING IT. AIM FOR MEDIUM-LEVEL CARBONATION, SIMILAR TO YOUR PALE ALE.

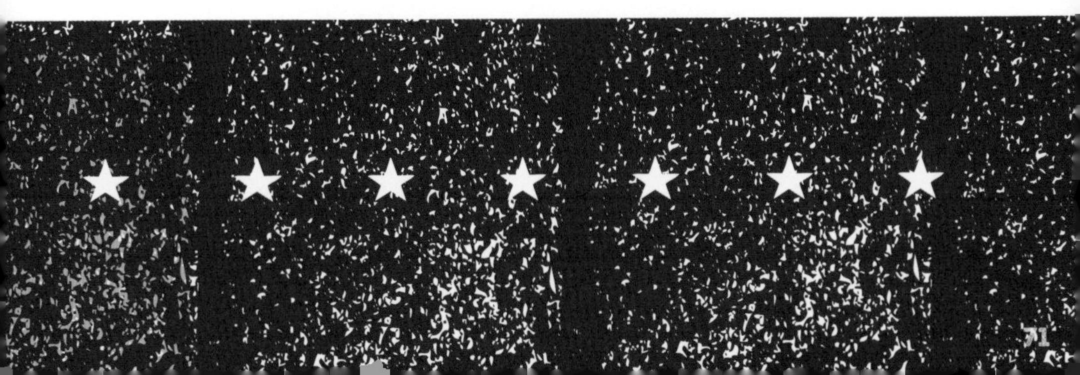

IHL – INTENSELY HOPPED LAGER

THIS RECIPE IS BASED ON ONE OF MY FAVOURITE COMMERCIAL BEERS – CAMDEN TOWN'S IHL (INDIA HELLS LAGER). IT'S A LAGER RECIPE BUT PACKED TO THE BRIM WITH HOPS LIKE AN AMERICAN IPA WOULD BE. HOPPY, JUICY, BALANCED AND REFRESHING.

ALTHOUGH IHL STANDS FOR INDIA HELLS LAGER, I THINK OF IT AS INTENSELY HOPPED LAGER. ALL THE FLAVOUR OF AN IPA, MINUS THE HEAVY SWEETNESS. IT'S BIG, BOLD AND JUICY – WITH A CRISP AND CLEAN LAGER FINISH.

TARGETS:
ORIGINAL GRAVITY: 1.060
FINAL GRAVITY: 1.010
ABV: 6.5%
COLOUR: 15 EBC

BOIL TIME: 60 MIN
BATCH SIZE: 19L

INGREDIENTS
GRAIN

4.8kg – Pilsner Malt
1.2kg – Munich Malt
300g – Carapils

HOPS

50g – Magnum – Boil 60 min
25g – Simcoe – Boil 10 min
25g – Chinook – Boil 10 min
25g – Mosaic – Boil 10 min
25g – Simcoe – Flame-out
25g – Chinook – Flame-out
25g – Mosaic – Flame-out
33g – Simcoe – Dry Hop
33g – Chinook – Dry Hop
33g – Mosaic – Dry Hop

YEAST

1 pack German Lager Yeast – ferment at 8° to 10°

After no more than 72 hours of dry hopping, transfer to a final vessel and try to chill for at least another two weeks.

TEASEL BEST BITTER

THIS IS A GREAT RECIPE THAT HAS A REGULAR PLACE IN THE STROUD BREWERY TAP ROOM. A CLASSIC BEST BITTER. FULL BODIED WITH A FRUITY MALT FLAVOUR AND THE INCLUSION OF BLACK PEPPER MAKES IT PARTICULARLY TASTY.

TARGETS:
ORIGINAL GRAVITY: 1.047
FINAL GRAVITY: 1.012
ABV: 4.9%
COLOUR: 18.6 EBC
BITTERNESS: 42.8 IBU

BOIL TIME: 60 MIN
BATCH SIZE: 19L

INGREDIENTS

GRAIN

3.4kg – Pale Ale Malt
350g – Caramalt
175g – Rolled oats
85g – Crystal Malt
20g – Chocolate Malt

HOPS

30g – Fuggles – Boil 60 min
20g – Hallertauer Blanc – Boil 5 min
25g – Crushed black peppercorns – Boil 5 min
15g – Motueka – Flame-out
15g – Simcoe – Flame-out
25g – Motueka – Dry Hop
25g – Simcoe – Dry Hop

YEAST

1 pack Safale S-04

RECIPE KINDLY GIVEN BY STROUD BREWERY

BROOKLYN SORACHI ACE

GARRETT OLIVER IS A BIT OF A LEGEND IN THE BREWING WORLD. HE'S BEEN BREWMASTER AT THE BROOKLYN BREWERY SINCE 1994 AS WELL AS BEING A HIGHLY REGARDED AUTHOR AND LECTURER ON ALL BEERY MATTERS. I'VE LOVED EVERY SINGLE BROOKLYN BREWERY BEER I'VE TASTED (AND THEY MAKE A LOT) BUT ONE OF MY FAVOURITES IS THEIR SORACHI ACE – PACKED FULL OF THE DELICIOUS SORACHI ACE HOP. I WAS PLEASED TO HEAR IT WAS ALSO ONE OF GARRETT'S FAVOURITE BEERS TO DRINK AND ABSOLUTELY STOKED WHEN HE GAVE ME HIS RECIPE TO BREW IT AT HOME.

SO, HERE IT IS FOLKS – A BIT TRICKIER THAN YOUR USUAL HOME-BREW RECIPE, BUT SEEING AS IT IS THE FAVOURED BREW OF ONE OF THE WORLD'S MOST AWESOME BREWERS THEN YOU OWE IT TO YOURSELF TO HAVE A CRACK AT MAKING IT.

TARGETS:
ORIGINAL GRAVITY: 1.062
FINAL GRAVITY: 1.008
ABV: 7.5%
COLOUR: 6 EBC
BITTERNESS: 36 IBU

BOIL TIME: 60 MIN
BATCH SIZE: 19L

INGREDIENTS

GRAIN
5kg – Pilsner Malt
450g – Corn Sugar

HOPS
17g – Sorachi Ace – Boil 60 min
17g – Sorachi Ace – Boil 30 min
150g – Sorachi Ace – Flame-out
60g – Sorachi Ace – Dry Hop for 5–7 days

YEAST
1 pack Wyeast 1214 (Belgian Ale) or White Labs 500 (Trappist Ale) Yeast

The difference in this recipe and others is we're going to start the steeping process a little lower and then gradually raise the temperature in stages. You can easily do this by firing up the burners on your hob for a minute or two – make sure you keep stirring!

Temperature steps are:
50°C for first 10 minutes.
Raise temperature to 63°C and hold for 60 minutes.
Raise mash temperature again to 67°C and hold for 15 minutes.
Finally raise to 75°C then remove the grain and start boiling.

TINY REBEL – URBAN IPA

THIS WAS ONE OF TINY REBEL BREWERY'S FIRST IPAS, KIND OF A CLASSIC AMERICAN EAST COAST IPA STYLE WITH RESINOUS, DANK AND SPICY FLAVOURS AND AROMAS.

TARGETS:
ORIGINAL GRAVITY: 1.051
FINAL GRAVITY: 1.010

BATCH SIZE, 20L

INGREDIENTS

GRAIN

3.6kg – Pale Malt
100g – Carapils
100g – Crystal 150
120g – Munich
100g – Wheat

HOPS

5g – Columbus – Boil 60 min
18g – Chinook – Boil 15 min
18g – Cascade – Boil 10 min
22g – Columbus – Boil 5 min
22g – Chinook – Boil 5 min
22g – Cascade Flame-out
50g – Chinook Dry Hop
50g – Columbus Dry Hop

YEAST

1 Pack WLP 007 or equivalent

PILS

THE TERM 'LAGER' REFERS TO THE WAY A BEER IS FERMENTED AND CONDITIONED IN A COLD TEMPERATURE, A PROCESS KNOWN AS 'LAGERING'. IDEALLY YOU CAN FIND SOMEWHERE IN YOUR HOUSE DURING THE WINTER THAT STAYS AT A CONSTANT 10°C – PERHAPS A BASEMENT OR SHED. IT'S AN IDEAL TIME TO BREW AND STOCK UP ON LAGERS READY FOR THE SUMMER.

IF NOT, YOU CAN STILL GET HALFWAY TO A DECENT PILSNER OR LAGER BY USING THE CORRECT STYLE OF MALT AND HOP, FERMENTING AS COOL AS YOU CAN FIND AND THEN CONDITIONING SOMEWHERE COLD FOR A GOOD FOUR WEEKS OR SO. THERE ARE SOME GOOD TIPS ON LOW-TECH LAGERING TECHNIQUES ON THE BASIC BREWING RADIO WEBSITE.

ORIGINAL GRAVITY: 1.044
FINAL GRAVITY: 1.009
ABV: 4.5%
COLOUR: 5.7 EBC
BITTERNESS: 24 IBU

BOIL TIME: 60 MIN
BATCH SIZE: 19L

INGREDIENTS

GRAIN

4.25kg – Lager or Pilsner Malt

HOPS

40g – Tettnang Hops – Boil 60 min
30g – Saaz Hops – Boil 30 min
15g – Saaz Hops – Boil 10 min
10g – Saaz Hops – Flame-out

YEAST

1 pack SafLager Yeast West European Lager

Brew the beer as usual and then ferment at 10°C.

Condition for at least four weeks at 10°C before bottling.

BREWERY TOURS

BRICK BREWERY

PECKHAM, LONDON

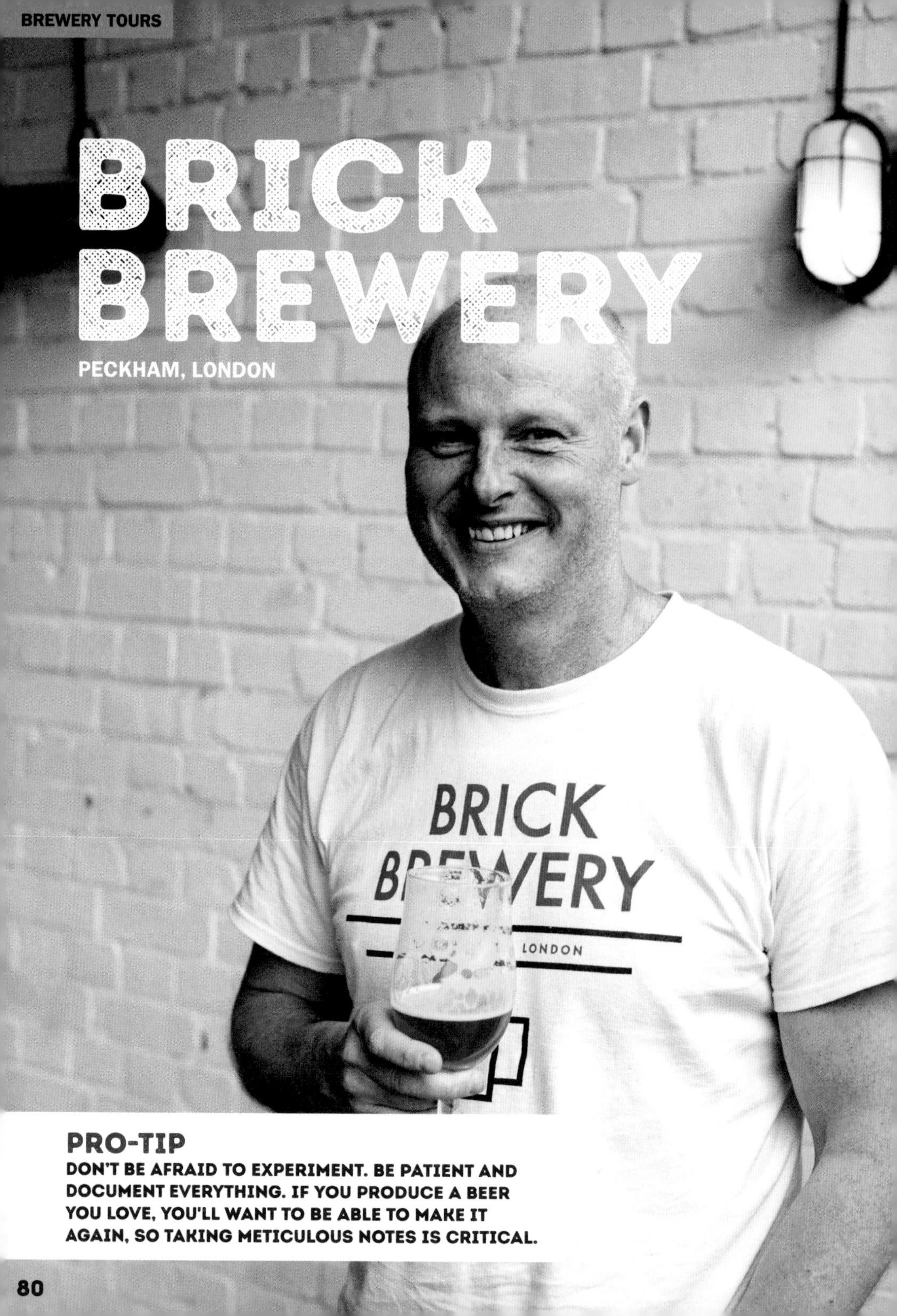

PRO-TIP

DON'T BE AFRAID TO EXPERIMENT. BE PATIENT AND DOCUMENT EVERYTHING. IF YOU PRODUCE A BEER YOU LOVE, YOU'LL WANT TO BE ABLE TO MAKE IT AGAIN, SO TAKING METICULOUS NOTES IS CRITICAL.

WE BUILD BEER!
IN FACT, WE'VE PRETTY MUCH BUILT EVERYTHING WITH OUR OWN HANDS.
OUR BEERS ARE UNCOMPROMISING.
WE CONSTANTLY STRIVE TO BREW BETTER BEER AND CHALLENGE THE PERCEPTION OF THE EVERYDAY DRINKER WITH OUR CORE RANGE BUT ALSO THE CRAFT-BEER DRINKER WITH OUR SEASONAL BEERS.

TELL ME YOUR NAME AND A BIT ABOUT YOURSELF.

Ian Stewart, born in York, and after a sixteen-year stint living in the US I moved to London. Having worked in marketing for twenty-two years I started brewing in my shed at the bottom of the garden. A few years later I had secured a railway arch in Peckham Rye and bought a 6bbl (barrel) brewhouse with two fermentation tanks.

HOW DID YOU GET INTO BREWING IN THE FIRST PLACE?

My wife bought me a Christmas gift – just add water and have beer in three weeks. I sat on it until she pressed me to make it. One summer afternoon I went to my shed and never looked back, buying progressively larger brewpots and fermentation buckets, finally ending up with a 100L mini-brewhouse and shelves full of beer.

WHAT WERE THE STEPS THAT LED YOU TO START A BREWERY?

I have always wanted to work for myself. I started a Mexican taco truck business selling at various markets around London. One July 4th private event for a group of Americans I sold maybe 200 tacos but 1000 bottles of Corona. The maths was simple; it was a no-brainer. Beer was the future. It took two years of planning and careful consideration but in October 2013 we brewed our very first commercial beer.

TALK US THROUGH YOUR BREWERY.

Brick is a 6bbl brewhouse so each time we brew we produce 1100 litres. Perfect amount for our original 1000L fermentation tanks. However, just one year in we bought three 2200L fermentation tanks, which means we need to brew twice to fill them. Then last year we bought four 3000L fermentation tanks, meaning it needs a triple brew to fill. Because of this we typically brew six times per week. The larger tanks were brought in to futureproof the business as this year we are looking to upscale again, moving production from our humble bursting-at-the-seams 1000sqft arch into a 6000sqft site. The site will be home to a new 3000L brewhouse and an additional eight fermentation tanks, taking our overall capacity to 44,000L (77,000 pints). Included in the expansion will be a new kegging line and a new canning line. We'll continue to produce our core range at greater scale but also introduce new seasonal beers as well as experiment with barrel ageing and sour beers.

WHAT'S YOUR FAVOURITE BEER TO BREW?

At the brewery we are very proud of all the beers we produce. We've worked very hard getting them to where they are today so I don't really have a favourite beer to brew. Our range is quite diverse and demonstrates our ability to produce consistent beers inspired by different regions of the world. Whether that's the hop forward Peckham Pale Ale or Pioneer IPA to the very fresh crisp Czech-style pilsner, Peckham Pils.

WHAT'S YOUR FAVOURITE BEER TO DRINK?

That all depends on the mood I'm in and what I'm eating. More and more I find myself reaching for a beer that pairs nicely with food. Our seasonal US Brown goes very nicely with pulled pork and the Pioneer IPA goes down a treat with spicy Thai food.

WHAT'S YOUR FAVOURITE THING ABOUT WHAT YOU DO?

We're in a pretty sociable industry and our taproom at the brewery continues to be a place where we get to meet not only our regulars but also people who are being introduced to craft beer. Our beers are accessible to anyone, making our taproom atmosphere all-inclusive and a relaxed place to enjoy a pint or two. We brew small-batch taproom exclusives – just 88 pints each time – and once it's gone it's gone. It's great to see people come just to try our latest ginger and pepper pale ale, or our cherry sour or an Earl Grey Mild. We get instant feedback and gratification for the brewers and the bar staff.

WHAT MAKES YOUR BEERS UNIQUE?

Our beers are uncompromising. We never settle for mediocrity so our beers use the freshest ingredients possible. You can't cut corners or scrimp on ingredients if you want the beer to taste exactly how you want it to taste. If a beer isn't at the level we think it should be it doesn't go to market.

TINY REBEL BREWERY

PRO-TIP

START BASIC, LIKE OUR CWTCH HOME-BREW KIT, WHICH IS A GREAT WAY FOR NEWBIES TO GET GOING. DON'T GET TOO BOGGED DOWN IN THE TECHNICAL STUFF STRAIGHT AWAY AND ENJOY IT!

WE BELIEVE BEER IS ABOUT HAVING FUN, AND THAT COMES ACROSS IN EVERYTHING WE DO. OUR BRAND IS BASED AROUND A SCRUFFY BEAR, WITH LOTS OF COLOUR AND LOTS OF JOKES IN OUR DESIGNS.

WE MAKE BEER FOR CASK, KEG, BOTTLE AND CAN AND FIRMLY BELIEVE THAT GOOD BEER IS GOOD BEER NO MATTER THE PACKAGE.

TELL ME YOUR NAME AND A BIT ABOUT YOURSELF.

We're Tiny Rebel, we're from Newport in Wales, we've won two Champion Beer of Wales trophies and one Champion Beer of Britain award since opening five years ago and we love everything from proper bitters to fruit sours.

HOW DID YOU GET INTO BREWING IN THE FIRST PLACE?

Gazz, our Head of Beery Stuff, grew up around home brewing. He was fascinated by his grandfather's bottles of ginger beer fermenting (and sometimes exploding) under the stairs. He helped out, learned as much as he could and then started home-brewing himself a little later. He got his brother-in-law and colleague Bradley interested in the hobby and they started brewing together on weekends.

WHAT WERE THE STEPS THAT LED YOU TO START A BREWERY?

Brad and Gazz home brewed in Brad's dad's garage for a few years and after realising that friends they were giving their beers to kept coming back for more, they started to wonder why they couldn't get beers of similar quality in pubs around us. A lot of research trips to pubs and a few versions of a business plan later, Tiny Rebel was born.

TALK US THROUGH YOUR BREWERY.

We brew on a dual-stream, 30bbl brewhouse which means we can brew twice pretty much simultaneously. At maximum capacity, we'll be able to brew on each twice a day. We're currently brewing ten times a week.

WHAT'S YOUR FAVOURITE BEER TO BREW?

Cwtch, which won Champion Beer of Britain in 2015, fills the brewery with caramelly malt and citrusy hop aromas. It's a super-drinkable ale and those smells just get the mouth watering.

WHAT'S YOUR FAVOURITE BEER TO DRINK?

Every member of the team will likely have a different answer, and probably a few of them! It totally depends on mood, surroundings and time of year. Dark winter nights are perfect with a pint of Dirty Stop Out, but a can of Clwb Tropicana is perfect when the shorts and sunglasses are out. All of our beers are favourites at one time or another!

WHAT'S YOUR FAVOURITE THING ABOUT WHAT YOU DO?

We love brewing new specials and seasonals. In 2016, we set ourselves a challenge and brewed thirty new beers in addition to our usual seasonals and our core range. We try out a lot of ideas that come from everywhere in the team – from brewers to our office staff and the guys in our bars!

WHAT MAKES YOUR BEERS UNIQUE?

The combination of perfectly balanced flavours, lots of experimentation, a sense of humour with our branding and pure-and-simple drinkability. Nobody does it like we do it!

DO YOU HAVE ANY BREWING PHILOSOPHIES THAT HAVE DRIVEN YOUR CHOICE IN WHAT BEER TO MAKE?

Give it a try! The worst that can happen is it won't be perfect. We love trying new things and we'll give anything a go. Every idea is worth thinking about!

WILD CARD BREWERY

PRO-TIP

DO EVERYTHING YOU CAN TO KEEP YOUR TEMPERATURES CONSISTENT – IT MAKES ALL THE DIFFERENCE.

WILD CARD BREWERY IS ALL ABOUT MAKING GREAT BEER AND GETTING IT OUT THERE. WE STARTED IT ON A SHOESTRING BUDGET – PARTIALLY FUNDED BY PERSONAL CREDIT CARDS, PARTIALLY FUNDED BY PEER-TO-PEER LENDING. WE JUST DECIDED TO HAVE A GO AND SEE WHAT HAPPENED.

WHEN WE MOVED TO OUR CURRENT SITE AT THE RAVENSWOOD INDUSTRIAL ESTATE IN WALTHAMSTOW, LONDON, WE DECIDED TO OPEN A BAR AT THE WEEKENDS, WHEN WE AREN'T BREWING. THAT'S BEEN GREAT BECAUSE IT HELPS US TO GET TO KNOW OUR CUSTOMERS AND FOR THEM TO GET TO KNOW US.

IT'S ALSO REALLY IMPORTANT TO US THAT WE DO THINGS RIGHT, AND THAT MEANS MAKING SURE NO ONE IS EMPLOYED ON ANYTHING LESS THAN THE LONDON LIVING WAGE.

TELL ME YOUR NAME AND A BIT ABOUT YOURSELF.

Wild Card Brewery is a micro-brewery based in Walthamstow, London. It was founded by Andrew Birkby and William Harris in 2012, starting out as cuckoo brewers, and moving to our current site in January 2014.

HOW DID YOU GET INTO BREWING IN THE FIRST PLACE?

We grew up in the Midlands surrounded by great breweries, so trying new beers and learning about how they are made has always been really exciting for us. We got into brewing at home when we were teenagers, so it has definitely been something we've been passionate about for a while.

WHAT WERE THE STEPS THAT LED YOU TO START A BREWERY?

We were getting better and better at making beer at home, and we liked the idea of working for ourselves, so we just figured we would give it a go. Our Head Brewer, Jaega Wise, had a background in engineering and mathematics, so her coming on board meant we could do things properly from the off as well.

TALK US THROUGH YOUR BREWERY.

Our brewery is on the smaller side, but we work the kit really hard. It's a six brewers barrel brew length, and we currently produce about 5000L per week. We brew all day every day – our tanks are never empty. We're actually moving the brewery to another site later this year, so we can have more space, and to keep up with the demand! It will also give us space in our tanks to do more specials, and more experimenting.

WHAT'S YOUR FAVOURITE BEER TO BREW?

Our Ace of Spades London Porter. Being in London, the feed water that comes out the tap is perfect to make it, so the beer practically brews itself.

WHAT'S YOUR FAVOURITE THING ABOUT WHAT YOU DO?

Interacting with customers and talking with people who are enthusiastic about beer, which is why our bar is so important to us.

WHAT MAKES YOUR BEERS UNIQUE?

We have a great balance between ABV and flavour – they are very drinkable but full of flavour.

DO YOU HAVE ANY BREWING PHILOSOPHIES THAT HAVE DRIVEN YOUR CHOICE IN WHAT BEER TO MAKE?

We want to make sure we cover the key styles – an IPA, a porter, a lager – and that we do them right. After that we can get a bit experimental.

JACK
Wild Card
BREWERY

WILD BEER BREWERY

PRO-TIP

AFTER YOU ARE CONFIDENT YOUR WORT IS CLEAN AND COOL, IT IS ALL ABOUT FERMENTATION, AND BECAUSE OF THAT, FOCUS ON TWO THINGS: A) CHOOSING YOUR YEAST SPECIFICALLY FOR YOUR DESIRED OUTCOME AND B) CREATING THE RIGHT ENVIRONMENT FOR THAT YEAST TO DO WHAT YOU WANT IT TO DO.

THE WILD BEER CO. IS A BREWERY OF TWO HALVES. ON ONE SIDE WE ARE STRIVING TO BREW, BARREL-AGE AND BLEND THE VERY BEST LIQUID THIS COUNTRY CAN OFFER. ON THE OTHER SIDE WE ARE PUSHING OURSELVES AND SEEKING TO BREW MARKET-LEADING MODERN BEER.

NOW FOR THE 'COIN' OR THE ESSENCE OF WHAT BINDS BOTH SIDES TOGETHER, WE ARE FOREVER SEEKING TO PUSH THE BOUNDARIES AND CHALLENGE PEOPLE'S PERCEPTIONS OF WHAT BEER CAN BE.

TELL ME YOUR NAME AND A BIT ABOUT YOURSELF.

My name is Brett Ellis. I am from California, and for my entire adult life I have been focusing on food and flavour development in either a kitchen or a brewery.

HOW DID YOU GET INTO BREWING IN THE FIRST PLACE?

I was not allowed to work for the first few months of living here and someone lent me some home-brewing equipment. Penniless, bored and creative, I started brewing.

WHAT WERE THE STEPS THAT LED YOU TO START A BREWERY?

1 – Discontent. Andrew (the co-founder) and I would always work up new concepts and ideas together, most of which did not suit the business we were working with at the time.
2 – Location. We live in Somerset, the home of 'wild fermentation' in Britain, as far as scrumpy and cider go.
3 – Foresight. There was a swell in the distance and we both knew that we wanted to be a part of the new wave that was coming. So we started paddling hard to catch it.

TALK US THROUGH YOUR BREWERY.

We run a very basic 30hl brewhouse which has one single-infusion mash tun and two steam-powered kettles. This allows us to run three brews a day in sixteen hours. One cellar is largely made up of conical fermenters and a few bright tanks; it also operates an Andritz centrifuge and Rolec HopnikECO for dry hopping. Our bottling line is from Italy and our canning line is from the USA. Cellar Two is made up of five foudres that are French oak. I bought them from two wineries in California.

WHAT'S YOUR FAVOURITE BEER TO BREW?

I love both Modus Operandi and Sleeping Lemons. Modus is a long-term beer that takes about a year to make and blend. Sleeping Lemons uses interesting fermentation techniques and is a solid session beer.

WHAT'S YOUR FAVOURITE BEER TO DRINK?

Again, I've got to answer with at least two – Modus Operandi for its complexity and length of flavour and aroma, and then maybe POGO as it is so crushable.

WHAT'S YOUR FAVOURITE THING ABOUT WHAT YOU DO?

Probably the dynamic nature of our portfolio and business, everything from lager to sour, and seeing the beer tribe grow while making the beers we do.

DO YOU HAVE ANY BREWING PHILOSOPHIES THAT HAVE DRIVEN YOUR CHOICE IN WHAT BEER TO MAKE?

Fermentation is everything. Focus on the details. But I don't pull my hair out over what I don't know as the yeast always knows best. If you brew with a yeast called brettanomyces then brew with fear. And one more . . . brewing is a team sport.

THE WILD BEER CO
CHRONOS
Lagered in Foudre
+ Brettanomyces
DRINK WILDLY DIFFERENT

MOOR BEER COMPANY

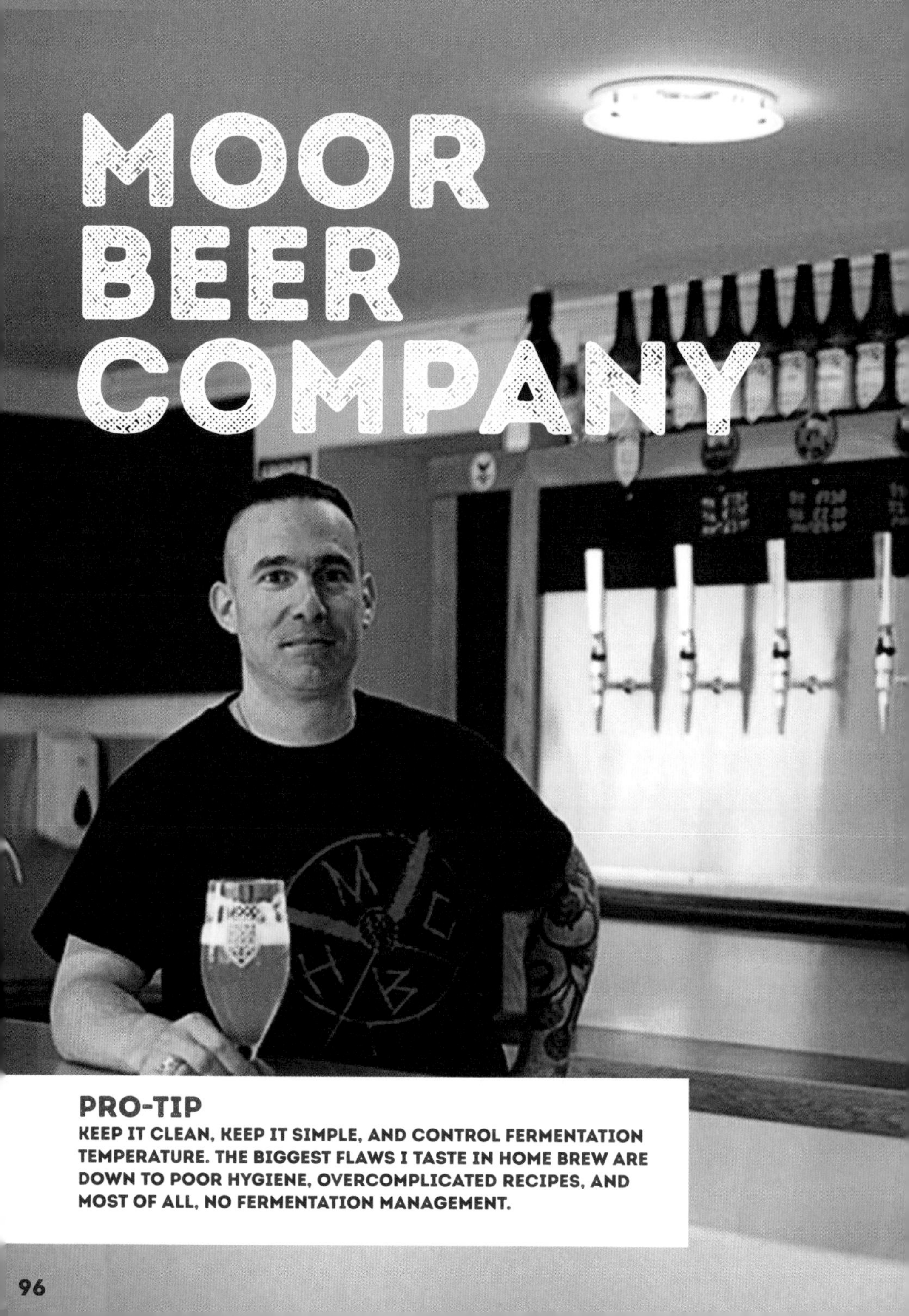

PRO-TIP

KEEP IT CLEAN, KEEP IT SIMPLE, AND CONTROL FERMENTATION TEMPERATURE. THE BIGGEST FLAWS I TASTE IN HOME BREW ARE DOWN TO POOR HYGIENE, OVERCOMPLICATED RECIPES, AND MOST OF ALL, NO FERMENTATION MANAGEMENT.

I PICKED UP WHAT UNDERPINS OUR PHILOSOPHY FROM THE VARIOUS PLACES I'VE LIVED AND TRAVELLED IN – HOPS AND FLAVOUR-FORWARD BEERS FROM CALIFORNIA, UNFINED (UNFILTERED) NATURALLY HAZY BEERS FROM GERMANY AND NATURAL CONDITIONING FROM THE UK.

WE MERGE THESE INTO WHAT WE CALL MODERN REAL ALE.

TELL ME YOUR NAME AND A BIT ABOUT YOURSELF.

Justin Hawke, owner and Head Brewer of Moor Beer Company. I'm originally from California, graduated from West Point, love England, pubs and real ale so moved here twenty years ago. Other passions are punk rock and *Star Wars* which you see incorporated into a lot our beers and events that we do.

HOW DID YOU GET INTO BREWING IN THE FIRST PLACE?

My TAC Officer at West Point was a home brewer and introduced me to the possibilities. A few years later my wife bought me a home-brew kit for Christmas and things got rolling from there. I'm sure she regrets that now!

WHAT WERE THE STEPS THAT LED YOU TO START A BREWERY?

In common with many home brewers, I had a dream of jacking in the day job and having my own brewery. There was previously a brewery called Moor that was started in the Levels & Moors area of Somerset by a dairy farmer. He wasn't a good brewer or businessman so eventually the brewery closed. We found out about it, bought the remains of the company, and started it as a completely new entity a couple years later. Everything needed to be replaced, including all the recipes, equipment, etc. In the end all we have from the original purchase is the name Moor Beer Company, the slogan Drink Moor Beer, and the name of one of our beers (Old Freddy Walker). In many ways we had an unnecessarily hard time taking this approach, but it taught us a lot of lessons and I think we're a much more solid company as a result.

TALK US THROUGH YOUR BREWERY.

We brew on a twenty-barrel (approximately 33hl) DME brewhouse with a single-infusion mash and steam-fired kettle – very much the typical British system. Everything in the brewhouse is manually controlled, so our beer is very much hand-made as opposed to rolling out of bed and checking the brew progress on your iPad. We use whole-leaf hops in the kettle and pellets in our cylindroconical unitanks. We do all our own packaging and have our own bottling line and canning line. It's a huge investment in time and equipment to take this approach, but we firmly believe it gives us more control over quality. Quality is hugely important to us, so we have an in-house laboratory as well, that is constantly being added to. Our latest addition is a yeast propagation system, which we're really excited to start using as it opens up a whole new world of possibilities for us to play with different yeasts and brew different beers.

WHAT'S YOUR FAVOURITE BEER TO DRINK?

My favourite beer to drink is one in perfect condition, be it a super-fresh IPA, a fantastically managed pint of traditional bitter, or a freshly tapped keller.

WHAT'S YOUR FAVOURITE BEER TO BREW?

Brewdays are pretty repetitive. Much like myself, many home brewers want their own breweries. And everyone wants to be a brewer these days! But people need to understand that brewing is not glamorous at all. It is mostly cleaning, with some set-up and take-down. It is exceedingly repetitive (some would say boring!), especially packaging. It is a 365-day job as you still need to do beer checks on holidays, often long hours in uncomfortable conditions for little pay. So I really try to impress upon aspiring brewers that they need to enter the trade with their eyes wide open. That being said, if you've got the right attitude then it's an amazing industry to work in and there is nothing else I would rather do (okay, maybe be in *Star Wars!*).

So when you ask what is our favourite beer to brew, they are largely all the same on the brewday. One of the exceptions, and one that we all enjoy taking part in even though it takes longer, is brewing our fresh hop beer, Envy. People wrongly expect fresh hop beers to be these massive hop-bombs, and they aren't at all. Fresh hops haven't been dried, so they are mostly water weight. Consequently you need to use a significantly larger quantity of hops in the brew, and those hops have a much subtler, grassier profile that don't give the hop resin character that some people may incorrectly expect. We are about an hour or so from the hop fields, so we drive up in the morning to get the freshly picked hops and repurpose the mash tun as a hop back. Packing in loads and loads of fresh hops is a lot of fun and the aroma is absolutely amazing. Cleaning out the mash tun afterwards is not so fun!

WHAT'S YOUR FAVOURITE THING ABOUT WHAT YOU GUYS DO?

We are small and family-owned so we can do whatever we want with the resources we have available. That means we've been able to do things like start the unfined beer movement, be the first to be accredited by CAMRA for real ale in a can, and work with hop researchers to develop new varieties. We also like getting involved with things like raising money for Hardcore Hits Cancer and travelling the world to collaborate with our brewing friends.

DO YOU HAVE ANY BREWING PHILOSOPHIES THAT HAVE DRIVEN YOUR CHOICE IN WHAT BEER TO MAKE?

As above, but also I really like the all-day pub/German beer hall drinking style, and that requires beers of relatively low alcohol so I deconstruct a lot of styles to make session-strength versions such as Black IPA and Hopefenweisse – beers that might have been 7–9% that we brew around 4.5% without sacrificing much flavour.

STROUD BREWERY

PRO-TIP

I WOULD ENCOURAGE OUR BREWERS TO SAY MORE ABOUT WHAT MOTIVATES THEM TO BREW, THEIR FAVOURITE BEERS AND THE CHALLENGES OF BREWING ORGANIC BEERS.

THE IDEA OF A BREWERY CAME TO ME WHILE WEEDING A CARROT FIELD WITH A FRIEND AND FELLOW BEER ENTHUSIAST, AND I WAS POSSESSED FROM THAT MOMENT ON.

STARTING A BREWERY COMBINED MY INTEREST IN BEER WITH MY WISH TO RUN A SMALL BUSINESS WITH CLOSE COMMUNITY CONNECTIONS. OUR FIRST BREW OF BUDDING WAS ON 29 MAY 2006.

ORIGINS OF THE BREWERY

I studied Marine Biology at Swansea University, continued a short career in conservation (English Nature and VSO in Nigeria at Yankari National Park) and then travelled for three to four years with some sponsorship from Guinness to study traditional brews of Africa, the drinks and the social settings they are drunk in. I returned and ended up working at the Soil Association in Bristol for its local food team. I managed a project promoting 'Community Supported Agriculture'. It was through this work I was introduced to Stroud, where my partner and I became founding members of Stroud Community Agriculture.

We started with a 5bbl brewery. By 2009 we had built up a loyal local pub customer base and were brewing at capacity. We moved into our existing building in early 2010 and our first brew was on 4 July 2010 – again, Budding. Our current brewery is a 20bbl plant and we produce about 12,000 pints a week. This is between casks that we sell to pubs, bottles that are distributed nationally, and beer sales through our own bar. The move was enabled with a European grant matched by a £100,000 loan from members of the local community.

My longstanding favourite beer of ours is Budding, but I have also become quite a fan of all our canned beers. I will always try new beers when I am out, but I am struggling to name a stand-out beer. There are so many great beers, different styles, different occasions, different moods . . . If I visit a good bottle shop I do look out for beers from http://brouwerijdemolen.nl/en/.

WHY ARE OUR BEERS UNIQUE?

We brew using malt grown on farms within approximately 15 miles of the brewery. Malted at Warminster Maltings, one of only three remaining traditional floor maltings. This craft malt is full of character and is the backbone of all our beers. We now brew almost entirely organic beers, so our enterprise supports organic farming in our local landscape where there is undisputed evidence that organic farming supports the highest biodiversity on farms. The standards also consider our brewing processes, the sustainability of the packaging we use, and staff welfare. We feel the organic standards are a good basis for producing great products.

What excites me most is our bar. Seeing people come together to drink our beers and have a great time. We can actually see interactions that are the stitches holding together the fabric of our local community.

CAMDEN BREWERY

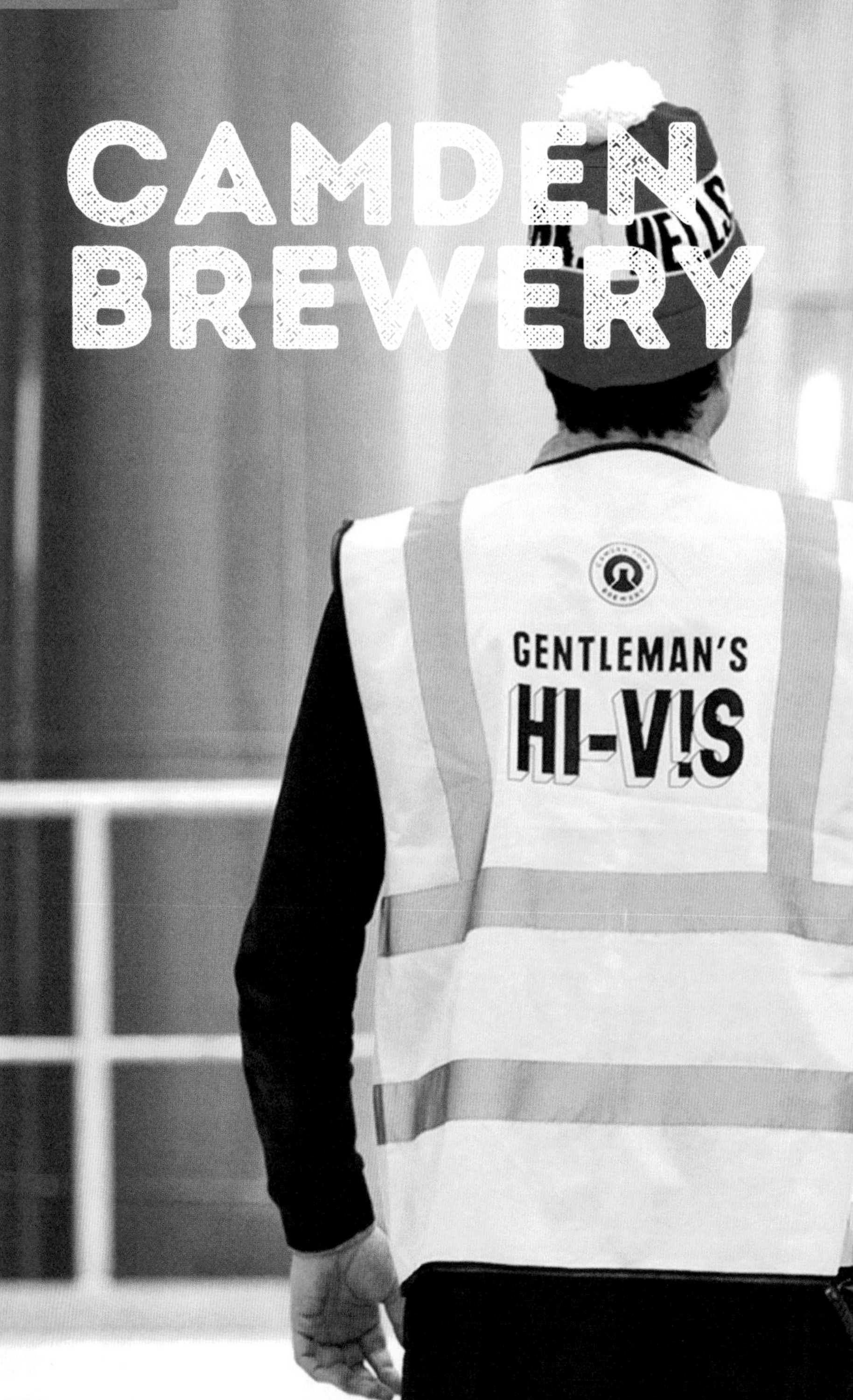

WE'RE ALWAYS LOOKING TO TAKE A STEP FORWARD WHEN IT COMES TO EVERYTHING FROM OUR BEERS, TO EVENTS, TO DESIGN, TO HOW WE COMMUNICATE OURSELVES. IT'S IMPORTANT TO MAKE YOUR OWN TRENDS WHILE TAKING INSPIRATION FROM OTHERS.

WHY DID YOU START A BREWERY?

I was brewing in the basement of my pub, the Horseshoe. I used to brew there two to three times per week; come in at 3 a.m., brew and cask in the same day, then hit the floor and wait tables. I'd play around with grains, hops, yeast: it was pretty primitive, but awesome fun. I loved it. Hands-on craft and exciting. But I loved lager. I'm Australian – I love beer with bubbles and I like beer served cold, full-flavoured beer, so there it was! I was going to make the best lager ever. We brewed our first beer, Camden Hells Lager, under the railway arches in Camden back in 2010 and we haven't stopped since.

WHAT'S YOUR GO-TO BEER?

It's got to be Hells Lager. I'm all about drinking big-flavoured beers, but what I look for is balance, which is why Hells is so appealing – it's crisp and refreshing and a beer that you can keep going back to.

WHAT'S YOUR FAVOURITE THING ABOUT WHAT YOU DO?

We take things up a notch, and go the extra mile and a half. We keep ourselves inspired by what's going on around us, in the beer world and further afield. We're always looking to take that step forward.

CAMDEN TOWN
BREWERY

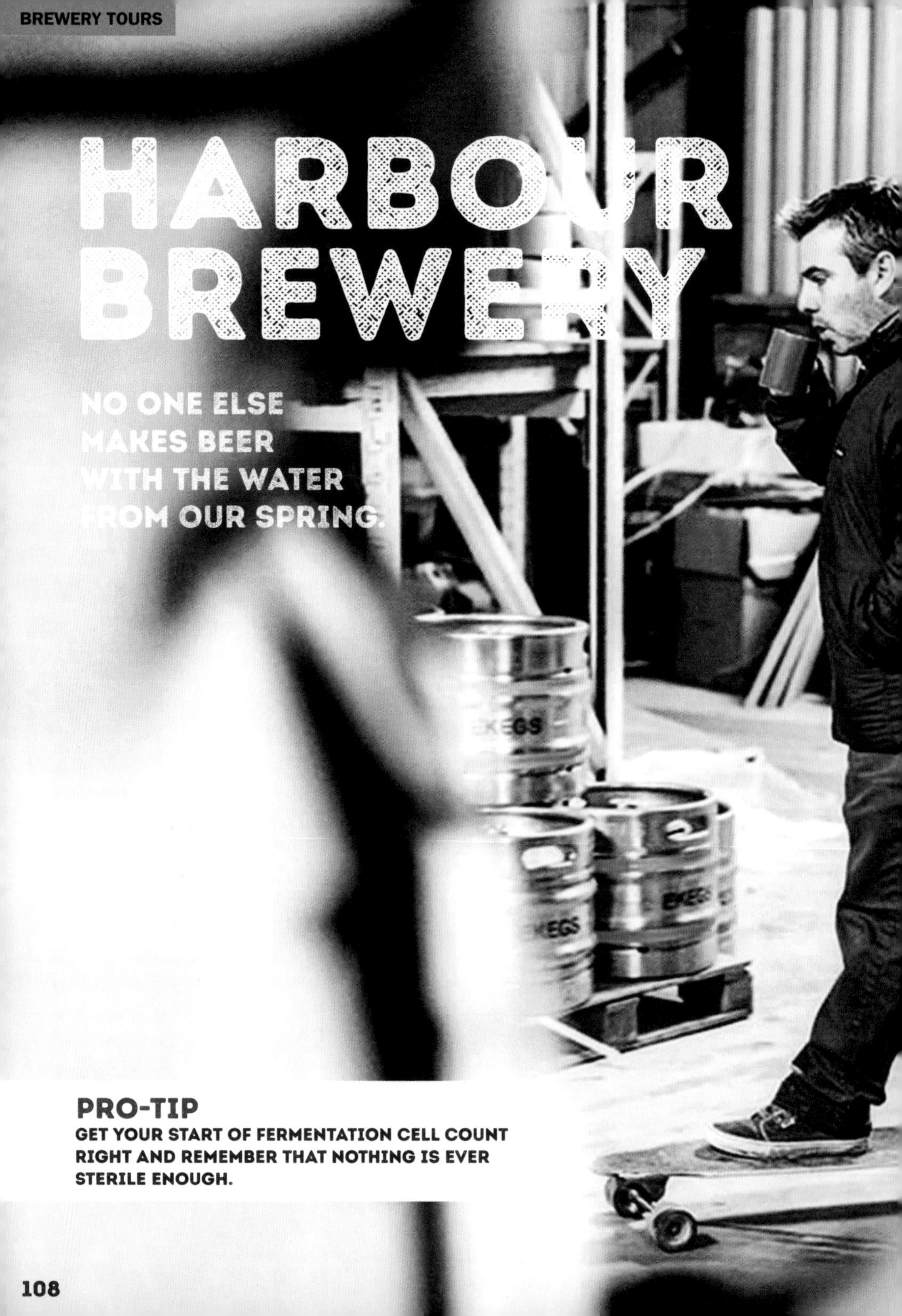

HARBOUR BREWERY

NO ONE ELSE MAKES BEER WITH THE WATER FROM OUR SPRING.

PRO-TIP

GET YOUR START OF FERMENTATION CELL COUNT RIGHT AND REMEMBER THAT NOTHING IS EVER STERILE ENOUGH.

HARBOUR WAS FOUNDED IN 2012. OUR AIM HAS ALWAYS BEEN TO MAKE INNOVATIVE AND FORWARD-THINKING BEERS THAT ARE BALANCED AND DRINKABLE. OUR BEERS AREN'T MEANT TO BE OVERLY CHALLENGING, THEY ARE JUST TO BE ENJOYED . . .

TELL ME YOUR NAME AND A BIT ABOUT YOURSELF.

Stuart Howe, Head Brewer of Harbour. I love beer and am obsessed with brewing:
Eddie Lofthouse, Founder, Harbour Brewing Company.

HOW DID YOU GET INTO BREWING IN THE FIRST PLACE?

I fell in love with beer at 18, went to evening classes to get the right A-levels and then did the Heriot Watt degree in brewing and distillery.
I have always enjoyed the variety of beers being made, and one of my friends was a brewer. One day we just decided to set up a brewery.

WHAT WERE THE STEPS THAT LED YOU TO START A BREWERY?

Not being able to easily buy the beers we wanted to drink down in Cornwall gave us the idea to start brewing commercially.

TALK US THROUGH YOUR BREWERY.

We have an American-built four-vessel brewhouse: mash, lauter and two kettle/whirlpools. A growing farm of 50hl FVs. Canning and bottling lines. And loads of cool stuff. We brew at least once a day on most days. It really depends on tanks: if a tank is empty we fill it!

WHAT'S YOUR FAVOURITE BEER TO BREW?

Lagers are hardest and give you a sense of achievement if they go well. Very hoppy beers are more relaxing because the hops dominate to the extent where minor issues aren't apparent.

Harbour Pale is my favourite beer to brew. Super-simple malt base, and bucketloads of Citra in the whirlpool. It smells amazing right through the whole process.

WHAT'S YOUR FAVOURITE BEER TO DRINK?

Everyone will say it depends on the time of day. I won't! I like elegant lagers and Belgian pales with mild bitterness.

Unlike Stuart, I like American-style pales and IPAs. Probably my favourite beer in the world is Deschutes Fresh Squeezed IPA – it's a Citra and Mosaic bomb. When it is fresh it is sublime.

WHAT'S YOUR FAVOURITE THING ABOUT WHAT YOU DO?

We have a team of people who know what they are doing, care about the beer and are not up their own arses.

The people. I get to travel around the world meeting brewers from all over the place. Pretty much everyone I meet in the industry is great to get on with and we often enjoy a beer or two together. I laugh a lot at the brewery – it is a fun place to be.

DO YOU HAVE ANY BREWING PHILOSOPHIES THAT HAVE DRIVEN YOUR CHOICE IN WHAT BEER TO MAKE?

Balance is critical. And by balance I don't mean balancing hops with more hops or mouth-excoriating bitterness with cloying sweetness!

Drinkability. I want to make beer people want to drink time and time again.

BUDWEISER BUDVAR BREWERY

PRO-TIP

DO YOUR READINGS FIRST. DEFINE YOUR OWN IDEA AND SUMMARISE WELL WHAT YOU WANT TO ACHIEVE.

AT BUDVAR WE ARE A TEAM OF APPROXIMATELY 600 PEOPLE WHO ARE PROUD OF THE BREWING HERITAGE AND TRADITION REACHING BACK 720 YEARS IN OUR REGION AND 122 YEARS IN OUR BREWERY.

TELL ME YOUR NAME AND A BIT ABOUT YOURSELF.

My name is Ales Dvorak, I'm 52 years old and I have been working at the Budweiser Budvar brewery since 1990. Besides beer brewing, which is my passion, I enjoy fishing and hunting. Another of my passions is history and warfare. I'm a reserve soldier of the Czech Army, where I'm ranked as a tanker. This is why I also support and enjoy tank beer.

HOW DID YOU GET INTO BREWING IN THE FIRST PLACE?

My subject at college was food industry, and I particularly liked beer brewing. It was at the college when I realised that I would like to become a real brewer. I felt proud to step up to this challenge to become part of the traditional and famous Czech beer industry.

WHAT WERE THE STEPS THAT LED YOU TO START A BREWERY?

After college and after the obligatory military service (one year back then) I started my first brewing job as a trainee in the Eggenberg brewery located in Ceský Krumlov. Back then the brewery was called Šumavský ležák – i.e. the lager of Šumava, the nearby bordering mountains between the Czech Republic, Austria and Germany, which was closed for forty years to the public owing to border protection during Communism.

TALK US THROUGH YOUR BREWERY.

We brew approximately 1.6 million hl per year, of which almost 900hl is intended for export. It is a very unique position among the Czech brewers concerning the global spread distribution in more than seventy countries worldwide.

We use traditional equipment. For brewing we use a double-decoction mashing process and we run two brew-sets, each of four vessels: the mash tun, mash pan, lauter tun and the wort copper.

One brew size is approximately 600hl and it takes about ten hours to get through. We brew daily, respecting the seasonal trends and equipment limits.

What makes our brewery different is the focus on long-term maturation of our beer. In our cellar we accommodate our lager for three months after its twelve days of main fermentation.

ALTOGETHER IT TAKES 102 DAYS TO BREW AND FULLY MATURE BUDWEISER BUDVAR ORIGINAL LAGER.

WHAT'S YOUR FAVOURITE BEER TO DRINK?

Besides our beers, I enjoy Belgian-style ales, Belgian lambic beers. I love British ales and occasionally I indulge in IPA-style beer, and of course I like the experimental projects with wide varieties of beers made with different US hops.

WHAT'S YOUR FAVOURITE BEER TO BREW?

I'm the father of our dark lager, and fathers tend to love their sons.

WHAT'S YOUR FAVOURITE THING ABOUT WHAT YOU GUYS DO?

The Czech Republic is well known for its lagers, however the beer awareness of Czechs is very limited in terms of beer styles. I love acting as an apostle of beer knowledge, sharing my experiences with the Czech audience. During my meetings with our customers I feel proud representing a brand which has been following the traditional production methods for generations. I feel I represent something that matters.

DO YOU HAVE ANY BREWING PHILOSOPHIES THAT HAVE DRIVEN YOUR CHOICE IN WHAT BEER TO MAKE?

Concerning Budvar it is pretty much given by the traditional raw materials and methods. However, sometimes, as an occasional home brewer, I try to figure out handy and easy methods to brew at home with the highest-quality result and reasonable time and energy input. I am not married, by the way!

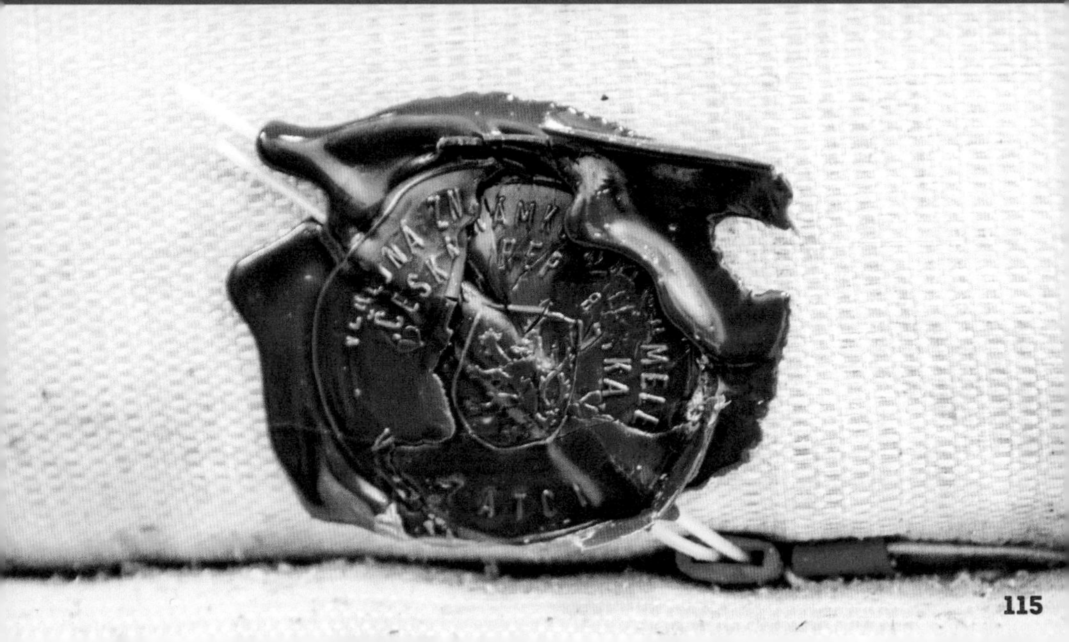

BROOKLYN BREWERY

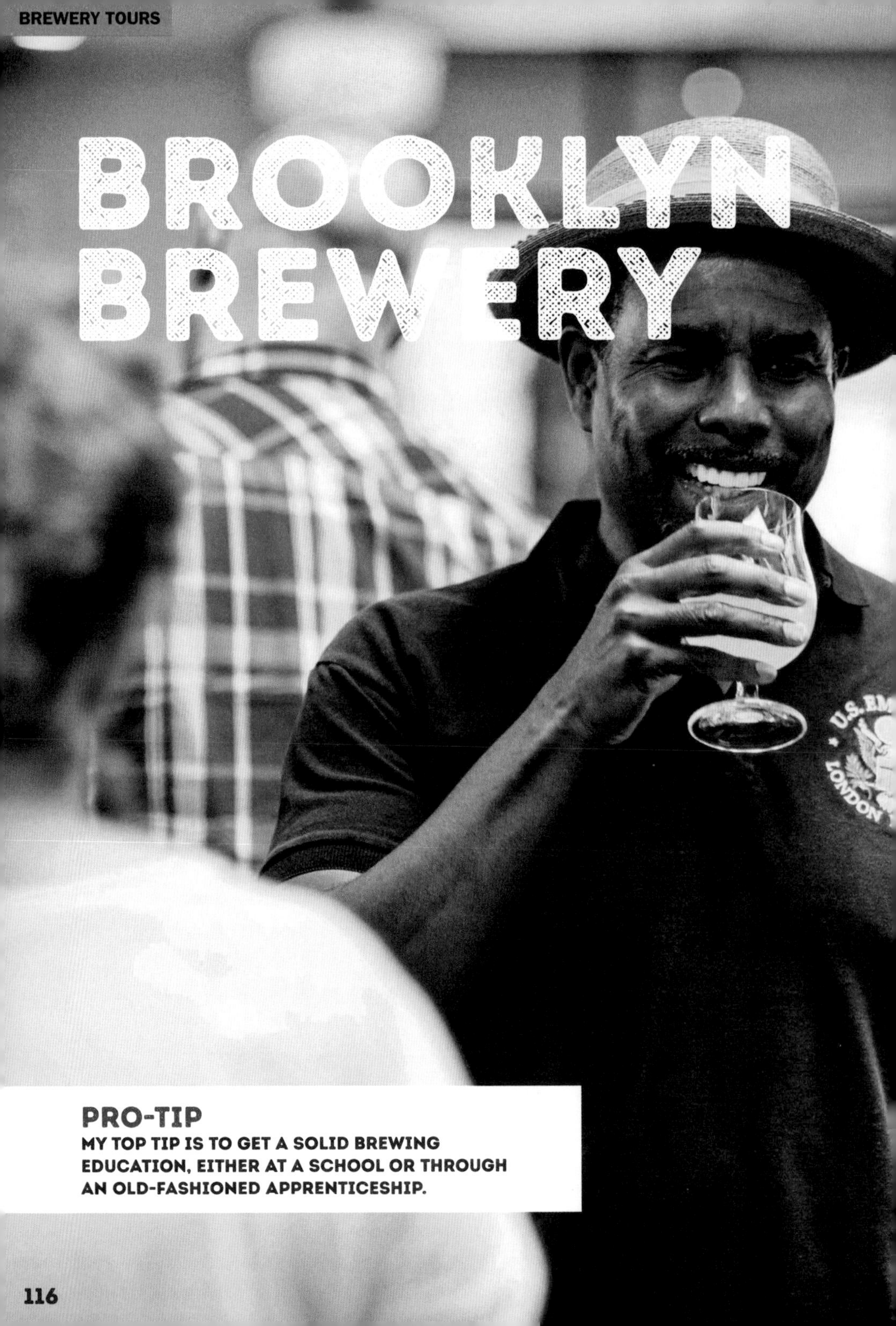

PRO-TIP

MY TOP TIP IS TO GET A SOLID BREWING EDUCATION, EITHER AT A SCHOOL OR THROUGH AN OLD-FASHIONED APPRENTICESHIP.

WE ARE BEST KNOWN TO FOLKS IN THE UK FOR OUR NOW-ICONIC BROOKLYN LAGER, WHICH IS THE DAY-TO-DAY TOP CRAFT BEER FOR NEW YORK CITY AND MANY OTHER FUN PLACES.

HOWEVER, IN THE US WE ARE KNOWN FOR THE THIRTY-FIVE DIFFERENT BEERS WE RELEASE EVERY YEAR AND FOR BEING ONE OF THE MOST INNOVATIVE BREWERIES IN THE WORLD.

TELL ME YOUR NAME AND A BIT ABOUT YOURSELF.

My name is Garrett Oliver. I'm Brewmaster of the Brooklyn Brewery, author of *The Brewmaster's Table*, and editor-in-chief of *The Oxford Companion to Beer* from Oxford University Press. My principal role at the brewery is the design of our beers, development of new beers, and the overall leadership of our brewing team.

HOW DID YOU GET INTO BREWING IN THE FIRST PLACE?

I moved to London from NYC in 1983. I stage-managed bands at the University of London Union (ULU) in Goodge Street, and put on shows by a lot of great bands. At the same time, though, I was falling in love with cask beer. When I moved back to the States I started home brewing. I went professional in 1989, as the apprentice to the former senior brewer for Samuel Smith's at Manhattan Brewing Company, an early brewpub.

WHAT WERE THE STEPS THAT LED YOU TO START A BREWERY?

I didn't found the brewery – Brooklyn Brewery started in 1988, and I joined in 1994. But I was here to build the current brewery and to build our portfolio of beers present and past from two beers to hundreds.

TALK US THROUGH YOUR BREWERY.

Our twenty-five-person brewing team is now making beer twenty-four hours per day, five days per week in Brooklyn. There are up to nine brews per day on a 60hl system. The system is very flexible and geared towards the wide range of beers that we make, including beers that come from our 2000+ oak barrels. We do a lot of bottle-conditioning and hand-work, so things are very busy all the time.

WHAT'S YOUR FAVOURITE BEER TO DRINK?

Some unknown, really delicious and inspiring new beer from some other brewery. Of our beers, well, you'll get a different answer every day. But I probably drink more Sorachi Ace and Bel Air Sour than anything else right now.

WHAT'S YOUR FAVOURITE BEER TO BREW?

When you make as many different beers as we do, that's difficult to say. My favourite thing recently was a beer that we made for the world-famous restaurateur Claus Meyer (he opened Noma in Copenhagen, for years considered the top restaurant in the world). We made a beer from left-over Danish rye bread from his baking operations at his NYC restaurant, Agern. They had frozen the bread; we thawed it and put it through a wood-chipper – almost 200kg of it. Then we added the kibbled bread to the mash and made it into a delicious beer!

WHAT'S YOUR FAVOURITE THING ABOUT WHAT YOU DO?

We are essentially a beer innovation engine. We invented a great many things that are becoming regular parts of the world of craft beer. We were among the first brewers in the world outside of Belgium to make saisons – we started in the 90s. We did the world's first beer collaboration brews, we have been making beers with the world's top chefs and restaurateurs, and we are making beers based on wild fermentations of wines and ciders. If you look at Serpent, the collaboration beer that we made with my good friends at Thornbridge Brewery and Tom Oliver's Cider & Perry, I can guarantee that you have never had a beer like that before. I love that beer, and bringing new ideas to beer is my favourite part of the work.

WHAT MAKES YOUR BEERS UNIQUE?

Our beers are boldly flavoured, but also smooth. Whether the beer is our 3.5% Saison, our new Bel Air Sour or our 14% barleywine, you will see the same idea – balance, structure, elegance and deliciousness.

DO YOU HAVE ANY BREWING PHILOSOPHIES THAT HAVE DRIVEN YOUR CHOICE IN WHAT BEER TO MAKE?

We only brew beers that we want to drink. If we don't want to drink it and we don't respect it, we're not going to make it. We also do everything we can to help out our fellow craft brewers – we must stick together to be strong and bring great new things to people. Fellowship is important. I always look to brew new beers that have 'talents' that our previous beers didn't have – beers that somehow bring something surprisingly delightful to beer drinkers.

WHAT PRO-TIPS CAN YOU OFFER THE HOME BREWER?

Professional brewing is to home brewing what being a restaurant chef is to being a home cook. That is to say that they are more different than they are alike. My top tip is to get a solid brewing education, either at a school or through an old-fashioned apprenticeship. Work for another brewery before you start your own. You will not find a single brewer who did that who isn't very, very glad he did.

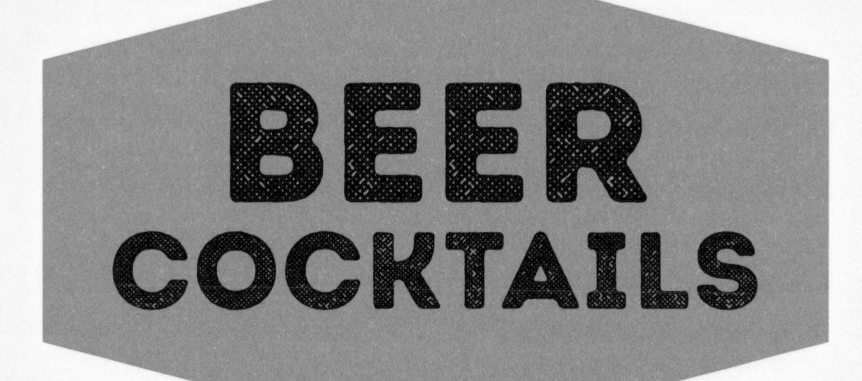
BEER
COCKTAILS

BOILERMAKER

THIS IS A BIT OF A HOOLIGAN OF A DRINK. THE SORT OF THING YOU ONLY REALLY ORDER AT THE BAR WHEN YOU'RE OUT WITH THE CREW AND HAVE A FEW IN ALREADY. WHENEVER YOU HEAR 'A ROUND OF BOILERMAKERS!!', YOU KNOW THINGS ARE ABOUT TO RAMP UP A NOTCH OR TWO. GRAB YOUR HAT AND HOLD TIGHT.

SERVES 1

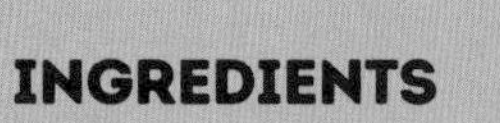

INGREDIENTS

Whisky
Beer

METHOD

Pour the whisky into a shot glass
Fill a pint glass halfway with beer
Drop the shot glass into the beer
Drink in one

BEERITA

WE ALL LOVE A MARGARITA, RIGHT? AND WE ALL LOVE BEER. SO IT'S CLEAR THAT IF WE COMBINE THE TWO, THEN WE'LL END UP WITH SOMETHING PRETTY AWESOME. REFRESHING, ZINGY, TEQUILA-EY. IT'S LIKE A CORONA WITH A HUNK OF LIME DIALLED UP TO 11.

THE STARTING POINT FOR A BEERITA IS AN AUTHENTIC MEXICAN LAGER. IT'S NOT SOMETHING I'D USUALLY CHOOSE OFF THE SHELF BUT IT WORKS BEAUTIFULLY WITH THIS. I LIKE DOS EQUIS IF YOU CAN FIND IT. HAS A BIT MORE CHARACTER THAN OTHERS.

INGREDIENTS

1 x 330ml bottle Mexican lager
37.5ml tequila
12.5ml Triple Sec
12.5ml fresh lime juice
Lime quarters to garnish
Salt

METHOD

Rub lime around the rim of a large glass and dip the rim into a saucer of salt

Shake the tequila, Triple Sec and lime juice together over ice and pour into the glass

Top up with Dos Equis

Chuck in a couple of lime wedges to garnish

OR FOR A BIT OF FUN AND FANCY PRESENTATION YOU CAN TOP UP TO 3/4 WITH THE DOS EQUIS AND THEN PUT THE WHOLE BOTTLE UPSIDE DOWN IN THE GLASS. AS YOU DRINK THE COCKTAIL (WITH A STRAW) THE BEER WILL KEEP TOPPING THE GLASS UP.

I HAVE NO IDEA WHY THIS IS CALLED THE ELVIS, BUT BEING A FAN OF ALL THINGS 'KING' RELATED, THIS HAS TO SIT IN MY TOP BEER-COCKTAIL LIST.

GIN COCKTAILS ARE ALL THE RAGE NOW. KICK THAT TONIC INTO TOUCH AND GET DOWN WITH SOME IPA TO MIX THINGS UP.

INGREDIENTS

Juice of two grapefruits
50ml gin
15ml elderflower liquor (or cordial)
1 bottle big and hoppy IPA with a grapefruit finish
Blood orange twist (for serving)

METHOD

Add grapefruit juice, gin and elderflower liquor to a cocktail shaker
Add ice and shake that thing
Strain over an ice-filled glass
Top up with IPA
Garnish with blood orange slice

STOUT FLOAT

THIS RECIPE COULD NOT BE EASIER. A SMALL HANDFUL OF INGREDIENTS IS ALL IT NEEDS TO MAKE YOU LOOK GOOD.

INGREDIENTS

1 BOTTLE OF GOOD STOUT
1 LARGE SCOOP OF VANILLA ICE CREAM
CHOCOLATE SAUCE
CARAMEL SAUCE

METHOD

POUR YOUR BOTTLE OF BEER INTO A FANCY GLASS

CAREFULLY ADD THE SCOOP OF ICE CREAM

DRIZZLE OVER THE SAUCES AND POP IN A STRAW

ONCE YOU HAVE THE INGREDIENTS IT'S EASY TO MAKE UP A BIG BATCH OF THE 'COCKTAIL MIX' AND THEN YOU CAN WORK THE PARTY WITH EASE. SLOSH OF COCKTAIL MIX, HAND OVER A BOTTLE OF BEER. BOOM! PARTY LEGEND.

INGREDIENTS

30ml spiced rum
30ml guava or agave nectar
30ml grapefruit juice
Mexican lager
Slice of lime to garnish

METHOD

Fill a glass with ice
Add the rum, nectar and grapefruit juice
Top with beer and stir
Add the lime to garnish

HELLFIRE

IF A COCKTAIL IS NAMED 'HELLFIRE' THEN IT HAD BETTER DAMN WELL LIVE UP TO IT.

THE TRICK HERE IS TO USE THE MOST FIERY, NOSE-BLASTING GINGER BEER YOU CAN FIND. MAKES A GREAT WINTER WARMER.

INGREDIENTS

30ml SPICED RUM
30ml GINGER BEER
2 DASHES TABASCO SAUCE
JUICE OF HALF A LIME
PILSNER

SERVES 1 DEVIL

METHOD

SQUEEZE THE LIME INTO A GLASS
FILL WITH ICE
ADD RUM, TABASCO AND GINGER BEER
FILL TO THE TOP WITH BEER

APEROL MIST

BELGIAN WHEAT BEERS HAVE A SLIGHTLY SWEET, CITRUSY SIDE TO THEM THAT COMPLEMENTS THE BITTER APEROL PERFECTLY.

A GREAT SUMMER GLUGGER.

INGREDIENTS

30ml Aperol
30ml fresh lemon juice
Belgian-style wheat beer (such as Hoegaarden)
Slice of blood orange to garnish

METHOD

Fill a large glass (pint size or so) with ice
Pour over the Aperol and lemon juice
Fill to the top with beer
Garnish with the blood orange

EVERYONE LOVES A MOJITO. SOMETHING ABOUT THAT MINTY, LIME ZESTY, REFRESHING HIT. THIS IS BETTER BECAUSE IT HAS BEER IN IT.

ALSO MAKES IT INTO A LOVELY LONG DRINK FOR THE SUMMER.

INGREDIENTS

30ml rum
10 mint leaves
1 tbsp sugar
30ml lime juice
200ml beer
Extra mint and slice of lime to garnish

METHOD

Muddle the mint with the sugar in a cocktail shaker
Add ice, lime juice and rum, then shake
Strain into ice-filled glass
Top up with beer and add more mint and a slice of lime to garnish

EVERYBODY LOVES A SUMMER PUNCH. HALF THE THRILL IS NOT KNOWING WHETHER IT'S GOING TO KNOCK YOU ON YOUR ASS IN TWO SLURPS, OR WHETHER IT'S SOMETHING YOU CAN GLUG AWAY ON ALL AFTERNOON. AND AS THE HOST, I FIND IT ENTERTAINING TO MAKE A PUNCH UNDETECTABLY POWERFUL AND WATCH THE CARNAGE COMMENCE.

NOTHING GETS A PARTY STARTED LIKE ROCKET-FUEL PUNCH.

THIS IS AN EASY-PEASY, KNOCK-TOGETHER SUMMER PUNCH. BUNG IT ALL IN A BOWL WITH A LADLE. FREESTYLE IT WITH THE FRUIT A LITTLE.

SERVES 10

INGREDIENTS

250g frozen raspberries
1L Belgian fruit beer – raspberry or cherry work well
1L raspberry lemonade
150ml vodka
Lemon and lime slices to garnish

METHOD

Mix all the ingredients together in a large bowl
Serve over ice
Sit back and watch

CAMPARI IPA

A DRINK FOR THOSE BITTER FOLK OUT THERE. DRY TO THE BONE.

INGREDIENTS

50ml Campari
IPA (India Pale Ale) Beer
Orange slices to garnish

METHOD

Chuck two or three ice cubes into a rocks glass
Pour over the Campari
Fill to the top with beer
Add a slice of orange to garnish

Caipbeerinha

CRAFT-BEER CONSULTANT AND BEER-ENTHUSIAST EZRA JOHNSON-GREENOUGH IS CREDITED WITH CREATING THE CAIPBEERINHA. I'VE ADDED IT TO THE BOOK BECAUSE

A) I LOVE A REGULAR CAIPIRINHA AND B) I LOVE THE NAME

INGREDIENTS

50ml Cachaça
2 lime quarters
2 lemon quarters
8 mint leaves
10ml agave nectar
30ml big, hoppy IPA

METHOD

Muddle the citrus and nectar in a cocktail shaker
Add the mint and Cachaça and shake with some ice
Pour into a glass with ice
Top with the IPA

MOST OF THIS BOOK IS ABOUT MAKING AND DRINKING BEER.

HERE'S A RECIPE FOR THE DAY AFTER YOU HAVE DONE ALL THAT.

IT'S LIKE A BEERY BLOODY MARY THAT TAKES ITS CLAIM TO FAME FROM BEING IN THE TOM CRUISE MOVIE *COCKTAIL*.

LET'S BE CLEAR – THIS IS A FUNCTIONAL DRINK, NOT A PRETTY ONE.

INGREDIENTS

30ml vodka
170ml tomato juice
350ml beer – something light and not too punchy
1 raw egg
Slice of blood orange to garnish

METHOD

Pour the vodka and tomato juice into a chilled glass
Add the beer
Crack the egg into the glass

COOKING
WITH
BEER

MUSSELS IN BEER BROTH

MUSSELS USUALLY GET COOKED IN WHITE WINE, BUT DIG THE BEER JUST AS MUCH.

INGREDIENTS

1kg mussels, cleaned and de-bearded (broken or opened ones discarded)
Splash of olive oil
4 sprigs of thyme
3 garlic cloves, minced
2 large shallots, chopped
Salt & pepper
250ml beer
Large knob of butter
2 tbsp freshly chopped parsley

METHOD

Heat the olive oil in a deep pot with a lid.

Add the shallots and garlic and cook for one minute.

Add the mussels to the pot.

Pour in the beer and bring to a simmer.

Cook, shaking the pan every now and again until all the mussels have opened and are cooked – around five minutes.

Transfer the mussels to a large bowl with a slotted spoon making sure you throw away any that have not opened.

Add the butter and parsley to the sauce and bring to a boil, whisking until it all combines before pouring over the mussels.

Garnish with a lemon wedge.

BELGIAN CHERRY BEER GLAZED PORK TENDERLOIN

PORK LOVES A BIT OF FRUIT. IF YOU THOUGHT PORK AND APPLES WERE THE WINNING COMBO THEN THINK AGAIN. THIS RECIPE DELIVERS A DELICIOUS, SWEET STICKY COATING TO THE SUCCULENT PORK LOIN WITH A DEPTH OF FLAVOUR AND BACKGROUND KICK OF CHERRY FROM THE BEER. FEEL FREE TO SWAP OUT THE CHERRY BEER FOR OTHER FRUIT BEERS. A HIT OF CHILLI ALSO WORKS WELL IN THIS RECIPE IF YOU LIKE THE HEAT.

IF YOU WANT TO GET FANCY, GARNISH WITH A FEW TINNED CHERRIES. THE ONES MADE FOR PIES, NOT COCKTAILS.

INGREDIENTS

4 pork tenderloins

For the glaze
200g dark brown sugar
100ml Belgian cherry beer
100ml soy sauce
100ml ketchup
50ml honey

METHOD

Season the loins with a generous pinch of salt and pepper.

Either set up your BBQ to cook at a medium direct heat, or set your oven to 180°C.

If using a BBQ, cook the loins as you would a steak, turning often, every five to six minutes.

When the pork tenderloins reach an internal temperature of 65°C brush on the glaze.

After five minutes remove from heat (or the oven) and allow to rest under loosely tented foil for another five minutes.

Slice thinly and drizzle with more glaze.

BEER-BRAISED BEEF RIB

SLOW-COOKED, SUCCULENT, STICKY, UNCTUOUS, FALL-OFF-THE-BONE BEEF.

INGREDIENTS

2 racks Jacob's Ladder (short ribs) beef – aim for 2 bones per person
Salt and pepper
200ml stout
200ml good beef stock

METHOD

Preheat the oven to 140°C.

Season the ribs generously with the salt and pepper.

Add the stout and beef stock to a deep baking tray and lay the beef ribs on top, bone side down. The beef should be part submerged in the liquid.

Cook for one hour uncovered, foil and put back for a further three hours.

Check the ribs – the meat should have risen up the bone a little and the texture should have softened.

Baste the ribs and place back in the oven, uncovered, for another thirty minutes.

Baste and check again for doneness by inserting a wooden skewer into the meat – it should slide in without resistance.

Sprinkle with some salt flakes before slicing carefully into individual bones to serve.

Serve with a big pile of mashed potatoes and a side of heavily buttered and peppered greens.

BUTTERMILK CHICKEN

THERE'S SOMETHING VERY AMERICAN-SOUNDING ABOUT BUTTERMILK-FRIED THINGS. MATCH UP YOUR CHICKEN WITH A BIG, HOPPY AMERICAN IPA.

BUTTERMILK WORKS TO TENDERISE THE CHICKEN AND ALSO KEEPS THE MEAT MOIST AND JUICY. IF YOU CAN'T FIND BUTTERMILK THEN USE RUNNY PLAIN YOGHURT INSTEAD.

INGREDIENTS

Approx 1kg chicken pieces – thighs drumsticks & wings
500ml buttermilk
1 batch JFC seasoning (see page 162)
1 batch beer batter (see page 140)
Vegetable oil

In a large bowl, or sealable plastic bag, combine the chicken and buttermilk until it's completely covered.

Leave in the fridge for at least four hours but preferably overnight.

Thoroughly drain the chicken and pat dry with kitchen roll.

Liberally season with the JFC seasoning.

Transfer the chicken to the batter, turning to coat all over.

Heat around an inch of vegetable or sunflower oil in a heavy-bottomed, deep pan until it reaches 180°C.

In batches so as not to crowd the pan, lift the chicken from the batter and shake off any excess, then fry until the internal temperature reaches 75°C.

Allow the oil to regain temperature in between each batch.

Serve with a blue cheese dip and a glass of something cold, wet and hoppy.

BEER BATTER

ONE OF THE BEST USES FOR BEER IN THE KITCHEN IS MAKING BEER BATTER. THERE ARE A COUPLE OF WAYS TO DO THIS: THE TRADITIONAL BEER BATTER LIKE YOU'D FIND IN TRADITIONAL FISH & CHIPS, OR YOU CAN FANCY IT UP A BIT AND MAKE A BEER TEMPURA. THE MAIN DIFFERENCE IS THAT TEMPURA REQUIRES A THINNER, LIGHTER BATTER.

TRADITIONAL BATTER

INGREDIENTS

125g plain flour
1 egg, beaten
350ml cold beer

METHOD

Whisk all the ingredients together in a big bowl to combine into a smooth batter. Allow to rest in the fridge for half an hour before you use it.

TEMPURA BATTER

INGREDIENTS

85g plain flour
1 tbsp cornflour
200ml ice cold, highly carbonated beer
Pinch of salt

METHOD

Whisk all the ingredients together in a big bowl and use immediately.

In either case, make sure your batter is kept cold and your oil is kept hot.

Don't overcrowd your pan. Fry in small batches or the oil temperature will drop and your food will soak it up rather than fry.

Suggested uses: Chunky fish – traditional/Fried pickles – tempura/Onion rings – traditional/Shrimp – tempura/Asparagus – tempura

THE BATTERS WORK WELL WITH DIFFERENT THINGS. BIG CHUNKY PIECES OF FISH GO WELL WITH THE THICKER, MORE ROBUST TRADITIONAL BATTER. LIGHTER BITS OF VEG AND PRAWNS WORK BEST WITH THE TEMPURA.

BEERY CROQUE MONSIEUR

A FIRM SUNDAY EVENING FAVOURITE IN OUR HOUSE, DEAD QUICK AND EASY TO MAKE, ESPECIALLY IF YOU HAVE THE BEERY CHEESE NACHO DIP IN THE FRIDGE. IF NOT, JUST TOP WITH MORE CHEESE OR A SIMPLE ROUX.

IT'S EASY TO MIX THIS UP WITH OTHER CHEESES. TRY USING TWO OR THREE DIFFERENT ONES. I LIKE TO KEEP A GOOD MELTY CHEESE IN THE MIX FOR THE SHEER FUN OF EATING IT.

INGREDIENTS

Two thick slices of white bread, crusts removed
50g butter, melted
Sliced Swiss cheese
Sliced ham
French mustard
Beer cheese nacho dip (see page 146)

METHOD

Liberally brush one side of each slice of bread with the melted butter.

Place under a medium grill until golden brown – just one side.

Remove and spread the unbuttered, untoasted, soft side with a thin layer of mustard.

Layer up the ham and cheese slices on top of one slice – toasted side down.

Add the other slice of bread – both toasted sides of the bread should now be on the outside.

Pop back under the grill for a few minutes until the cheese starts to melt – keep flipping it if it starts to over-brown.

Finally, spread a thick layer of the nacho cheese sauce on top of the sandwich and stick it back under the grill until the sauce starts to go brown and bubble.

Beast mode Cheese Toastie.

BEER-CAN CHICKEN

A GREAT WAY TO COOK CHICKEN IN THE BBQ (OR IN THE OVEN FOR THAT MATTER). CRISPY ON THE OUTSIDE, SUCCULENT AND MOIST ON THE INSIDE.

THIS METHOD NOT ONLY INFUSES THE CHICKEN WITH A LOVELY BEERY AROMA, IT ALSO LOOKS HILARIOUS TOO. NOT A PARTICULARLY NOBLE WAY FOR THE OLD BIRD TO GO, BUT SHE SURE AS HELL TASTES GOOD.

INGREDIENTS

1 chicken
2 cans of beer
Salt & pepper

METHOD

Season the chicken with a generous pinch of salt and pepper.

Either set up your BBQ to cook at a medium indirect heat, or set your oven to 180°C.

Open a can of beer and drink a third of it.

Position the chicken legs down, so the can sticks up the jacksie and the bird stands upright.

If cooking in the oven, place in a baking tray and bung in the oven.

If cooking on a BBQ, cook indirectly, rotating every twenty minutes or so, ensuring even cooking.

While the chicken cooks, drink your second beer.

Cook until the internal temperature reaches 75°C or the juices run clear when stabbed in the thickest part with a skewer.

SPAGHETTI & MEATBALLS

THIS MEAL IS A HUG IN A BOWL. FEEL FREE TO DITCH THE PASTA AND STUFF A BIG FAT SUB ROLL FULL OF THESE MEATBALLS INSTEAD.

INGREDIENTS

For the meatballs
500g minced beef
500g minced pork (or good sausages squeezed out of their skins)
200g breadcrumbs
2 garlic cloves, crushed
50g finely grated Parmesan
1 tbsp mixed dried Italian herbs – basil, oregano, parsley etc.
1 tsp each salt & black pepper
2 eggs, beaten

For the sauce
2 tbsp olive oil
2 garlic cloves, crushed
2 tbsp tomato paste
2 litres passata
Generous pinch of salt and pepper
250ml light Italian beer such as Peroni or Moretti
Fresh basil to garnish

Plus around 75–100g of spaghetti or other pasta of your choice per person

METHOD

By hand, mix together all the meatball ingredients in a big bowl until everything is evenly distributed. Form the mixture into meatballs around an inch in diameter.

Either place in a roasting tin and roast in a hot oven until brown, or fry in a little olive oil until browned. Set the meatballs aside while you make the sauce.

Gently heat the crushed garlic in the olive oil in a heavy-bottomed pan without allowing it to brown.

Once the garlic has softened, add the tomato paste and stir well.

Now add the passata, beer and seasoning.

Bring to the boil and add the meatballs to the pan.

Cover and simmer very gently for around forty-five minutes until the tomatoes and the oil start to separate a little.

Cook the pasta according to the instructions on the pack.

Around a minute before the pasta is fully cooked, add it to the sauce and meatballs along with a few splashes (maybe 100ml) of the pasta cooking water.

Stir well and simmer for a further thirty seconds.

Garnish with fresh basil leaves.

SERVES
8

BEER CHEESE NACHO DIP

CAN YOU BELIEVE IT? BEER CHEESE DIP IS ACTUALLY A THING. IT'S ALL YOUR DREAMS COME TRUE. WHAT COULD BE BETTER FOR DIPPING TORTILLA CHIPS OR SMOTHERING NACHOS WHILE YOU KNOCK BACK A FEW COLD ONES?

INGREDIENTS

2 tbsp unsalted butter
3 tbsp all-purpose flour
175ml milk – use in thirds
125ml beer (nothing hoppy)
1 tsp Dijon mustard
¼ tsp salt
⅛ tsp cayenne pepper
350g grated Cheddar cheese

METHOD

Melt the butter in a thick-bottomed pan over a medium heat.

Add the flour, and continue to stir until it's the colour and consistency of cookie dough.

Take off the heat and add one third of your milk, stirring constantly until smooth.

Now add another third of milk, put back on the heat and keep stirring until smooth. Finally, do the same with the last third. Now add the remaining ingredients.

Keep heating and stirring until the sauce starts to gently bubble, then remove from the heat.

Pour into a bowl, press the surface with clingfilm and allow to cool.

BEERY NACHOS

**NACHO, NACHO MAN.
I WANNA BE A NACHO MAN.**

YOU CAN SPICE THIS UP A BIT IF YOU LIKE BY ADDING SOME FRESHLY CHOPPED CHILLIES TO EACH LAYER OR A SPRINKLE OF CAYENNE PEPPER ON THE CHIPS.

INGREDIENTS

Big bag of nacho chips
Beer cheese nacho dip (see page 146)
Jar of jalapeño peppers
Grated mild cheese such as Monterey Jack or mild Cheddar
Refried beans

METHOD

Preheat the oven to around 180°C.

Tip a layer of tortilla chips into a baking tray – not too thick, but make sure you can't see the bottom of the tray.

Sprinkle over jalapeños along with dollops of refried beans and nacho sauce.

Top with a layer of grated cheese.

Bang in the hot oven for five minutes until everything starts to melt.

Remove from the oven and repeat the steps all over again – more chips, more cheese sauce, more beans.

Back in the oven for another five minutes.

Top with more jalapeños and serve hot.

BEER DRINKIN' SOFT PRETZELS

THESE GERMAN-STYLE CHEWY, SOFT PRETZELS ARE GREAT TO MUNCH ON WHILE SINKING A FEW BEERS. ORIGINATING IN BAVARIA AND SERVED WITH THE LOCAL BEERS, THEY GO PARTICULARLY WELL WITH A WEISSBEER OR THREE.

IT'S A PRETTY SIMPLE DOUGH THAT GETS DUNKED IN WATER BRIEFLY TO ALLOW A SKIN TO FORM BEFORE THE PRETZELS ARE BAKED.

THE TRICKY BIT IS SHAPING THEM. BUT THEN IF IT'S JUST YOU AND YOUR FAMILY OR MATES EATING THEM, DOES IT REALLY MATTER IF THEY LOOK A BIT LUMPY AND BUMPY? THEY TASTE GREAT EITHER WAY. YOU MAY WANT TO GOOGLE SOME YOUTUBE INSTRUCTIONS FOR SHAPING GERMAN PRETZELS.

YES, I KNOW THERE'S NO BEER IN THE RECIPE, BUT THEY TASTE SO GOOD.

INGREDIENTS

For the dough

1kg plain white flour
150ml warm milk
370ml warm water
75g butter (unsalted)
1 tbsp malt extract (liquid or dried, or brown sugar)
2 tsp fast-action dried yeast
2 tbsp salt

For the dip

1L water
3 tbsp baking soda

METHOD

Make the dough as if you were making a bread dough. Easiest if you have a food mixer with a dough hook – just sling all the ingredients in and mix for five minutes. By hand, mix and knead until you have a soft dough.

Allow to prove for a good hour until it has doubled in size.

Knock the dough back and form into small batons ready to shape. You can make twists or figure-of-eight shapes if you prefer.

Leave for a further thirty to forty minutes for the second rise. Again, they should double in size. And you want to see a skin start to form.

Bring your water to the boil and add the baking soda.

Very carefully, lower the pretzels one by one into the pot of boiling water for about three to four seconds. Literally just a quick dip and they should float back up again. Hook them out with a slotted spoon.

Sprinkle each pretzel with rock salt and make a slash in the fattest knot part to allow it to expand in the oven.

Bake at 200°C for about fifteen minutes until beautiful and toasty brown.

LINGUINE WITH CLAMS, CHILLI & IPA

ITALY HAS A THRIVING CRAFT BEER SCENE AND I AM LUCKY THAT I AM ABLE TO GET OVER THERE EACH YEAR. THE TRIP IS NOT COMPLETE WITHOUT CHOWING DOWN A BOWLFUL OF LINGUINE AND CLAMS WHILE OVERLOOKING THE SEA.

INGREDIENTS

60–70 clams
250g linguine
3 tbsp good extra virgin olive oil
1 finely sliced red chilli, or more if you like it hot
2 finely chopped garlic cloves
150ml IPA
Knob of butter
1 bunch of chopped flat leaf parsley
Generous grind of black pepper

METHOD

Make sure the clams are thoroughly clean by soaking them in a big bowl of cold water for a couple of hours, then rinse thoroughly.

Bring a large stock pot of water to the boil and salt generously.

Add the linguine and cook to just under al-dente – probably a minute or two less than it says on the packet. As the pasta cooks, heat a good couple of glugs of oil in a wide pan over a medium heat.

Add the garlic and chilli to the oil and cook for a minute or so.

Now add the clams and close the lid. Allow to cook, shaking intermittently, for another thirty seconds.

Lift the lid, discard any clams that have not opened, then pour in the IPA. Crank up the heat and simmer the beer down a little.

Turn the heat back down, add the drained pasta along with a 100ml splash of the cooking water, and continue cooking for a further thirty seconds.

Add the knob of butter and parsley and mix well.

Serve alongside cold glasses of the same IPA you used to cook with.

IN MY EXPERIENCE IT IS THE SMALLER, FAMILY PLACES RIGHT ON THE BEACH WITH PLASTIC CHAIRS AND PAPER TABLECLOTHS, AND DINERS WITH BARE SANDY FEET AND DRIPPING SWIMSHORTS, THAT TEND TO SERVE UP THE BEST SEAFOOD.

THIS IS USUALLY COOKED DOWN WITH A GLASS OF LOCAL WHITE WINE, BUT A CITRUSY, BRIGHT IPA WORKS JUST AS WELL.

DRUNKEN IRISH SODA BREAD

BAKING BREAD IS A REAL DELIGHT. BOTH THAT AND BREWING BEER GIVE YOUR HOME AN INCREDIBLE SMELL, ALTHOUGH MY WIFE WOULD ARGUE THE AROMA OF BAKING BREAD WINS. IN FACT, MANY OF THE INGREDIENTS ARE SIMILAR – CRUSHED GRAINS, WATER, YEAST.

WHILE I LOVE TO MAKE BREAD WITH YEAST, THIS IS A QUICK YET HEARTY LOAF YOU CAN BE EATING WITHIN THE HOUR.

INGREDIENTS

250g wholewheat flour
250g plain white flour
1 tsp salt
1 tsp bicarbonate of soda
25ml olive oil
250ml buttermilk or natural yoghurt
250ml malty, dark Irish ale

METHOD

Mix the dry ingredients together.

Add the wet ingredients and mix quickly with a fork until combined.

Very gently bring it together with your hands and shape it into a loaf on some baking paper.

Score the bread then place in a pre-heated oven at 180°C for around half an hour.

You'll know it is done when the bottom sounds hollow when tapped.

SALT BEER CARAMEL

SALT CARAMEL IS THE BIG THING AT THE MOMENT. THAT PERFECT SAVOURY/SWEET COMBO HITS THE SPOT EVERY TIME. THIS IS EVEN BETTER BECAUSE IT IS BEERY . . .

INGREDIENTS

400g brown sugar
250g butter
250ml golden syrup
1 small can of condensed milk
1 tsp vanilla extract
350ml bottle of (not hoppy) beer
Maldon sea salt

METHOD

Melt the butter and brown sugar together in a medium saucepan over a medium heat.

Slowly add the beer, syrup and condensed milk while stirring constantly.

Gently heat (still stirring) until the temperature of the mixture hits 115°C.

Take the mixture off the heat and stir in the vanilla extract.

Pour the mixture into a 20x30cm pan lined with parchment paper.

Sprinkle Maldon sea salt on top of the hot caramel. Let the caramel set until firm enough to cut.

STOUT CHOCOLATE CAKE

A BEERY CAKE. WHAT COULD BE BETTER? THIS ONE IS EVEN MORE FUN AS IT ACTUALLY LOOKS LIKE A PINT OF STOUT.

INGREDIENTS

For the cake
250ml stout (preferably your own home-brewed one)
250g unsalted butter, cubed
80g cocoa powder
400g caster sugar
2 eggs
1 tsp vanilla extract
140ml soured cream
280g plain flour
2½ tsp bicarbonate of soda

For the topping
125ml double cream
50g unsalted butter, softened
300g icing sugar
125g full-fat cream cheese

METHOD

Preheat the oven to 180°C.

Gently heat the stout and the butter in a big pan (don't boil).

Mix the cocoa and sugar and then whisk into the hot beer and butter pan.

Beat the eggs, vanilla extract and soured cream then also add to the pan.

Mix the flour and bicarbonate of soda and whisk into the pan a little at a time to avoid lumps.

Pour the batter into a greased cake tin and bake for around forty minutes until a skewer comes out clean.

While the cake is in the oven, beat the double cream and cream cheese together then gradually mix in the icing sugar to a thick paste.

Once the cake has cooled, spread the topping on top so it resembles the head on a pint of stout.

PAELLA

COMMUNAL DINING IS A WONDERFUL THING. IN SPAIN, FRIENDS AND FAMILIES WILL BRING INGREDIENTS AND ALL PITCH IN TOGETHER TO COOK UP A WONDERFUL LUNCH IN A CRAZY-BIG PAELLA DISH THAT FEEDS DOZENS.

INGREDIENTS

1 large onion, finely chopped
A generous glug (4–5 tbsp) of olive oil
2 garlic cloves, crushed
½ can chopped tomatoes
½ tsp sugar
Generous pinch of salt
1 tsp sweet paprika
1 tsp turmeric
A good pinch of saffron threads
2 cups (400g) Spanish paella rice
1 semi-cured chorizo (about 150g), sliced
2 chicken breasts, cut into chunks
750ml chicken stock, plus more if needed
250ml light Spanish beer – Cruzcampo, Mahou or Estrella Damm
½ tin roasted red peppers
12 langoustines
16 mussels, scrubbed and cleaned
32 clams – cleaned as on page 150

METHOD

Ideally cook this in a really wide paella pan, or get your biggest widest frying pan with a lid (but this may be a squeeze so halve the recipe!).

Fry the onion and garlic in the oil until it starts to colour, then add the chopped tomatoes.

Add the salt, sugar, paprika, turmeric and saffron then stir well.

Add the chicken and chorizo and cook for a further few minutes until the chicken starts to brown and the chorizo starts to melt.

Add the rice and stir well.

Add the stock and beer.

Add the roasted peppers.

Stir again and spread everything out in the pan, then cover and cook over a low heat for twenty minutes.

Stir well and spread the rice out evenly in the pan (do not stir again).

Cover and cook the rice over a low heat.

After ten to fifteen minutes, place the langoustines and mussels on top.

Five minutes later, add the clams.

If the rice looks too dry, you can add another splash of stock or beer.

Allow the seafood to cook for a further ten minutes and then remove from the heat and rest for five minutes before serving.

SERVES 6

THERE IS NO SINGLE PAELLA RECIPE REALLY. THE DISH ORIGINATES IN VALENCIA, WHERE MUCH OF SPAIN'S RICE IS GROWN, AND LIKE MANY TRADITIONAL AND DELICIOUS MEDITERRANEAN DISHES IT WAS MADE WITH THE INGREDIENTS LOCAL PEOPLE HAD TO HAND. BEING A COASTAL AREA, THAT OF COURSE MEANT LOTS OF SHELLFISH AND CHUNKS OF FISH, BUT THEY'D ALSO CHUCK IN PRETTY MUCH ANYTHING THEY GOT THEIR HANDS ON THAT DAY – CHICKENS, FROGS' LEGS, SAUSAGES ETC.

I LIKE TO FREESTYLE THIS RECIPE IN A SIMILAR WAY – USE A BIT OF WHAT YOU FANCY THAT DAY.

BREWED IN NEW YORK
BROOKLYN
BRAND
SUMMER ALE
12 FLUID OUNCES

BROOKLYN
BREWERY

BEER-SPRITZED CHICKEN SKEWERS WITH LIME WEDGES

INGREDIENTS

4 large chicken breasts, skin removed
1 Mexican beer

Chicken seasoning
2 tbsp sea salt
2 tbsp brown sugar
2 tbsp paprika
1 tbsp garlic salt
1 tbsp celery salt
2 tsp chilli powder
1 tsp black pepper

METHOD

Slice the chicken breasts lengthways into thin strips – around 0.5cm.

Thread the strips concertina fashion onto metal skewers then dust with the chicken seasoning.

Set your BBQ up to cook directly at a medium heat – 160°C or so.

Grill the skewers, turning often for around fifteen minutes, spritzing the chicken with beer every few minutes, until golden brown and cooked through (internal temperature 70°C).

Squeeze fresh lime over the skewers just before you remove from the grill.

Serve with more fresh lime wedges.

THIS RECIPE DOES NOT CONTAIN BEER, BUT, LIKE FRED & GINGER, WINGS & COLD BEER ARE A MATCH MADE IN HEAVEN.

INGREDIENTS

24 chicken wings (12 wings, split in half)
1 tbsp of JFC seasoning (see page 162)
150ml Frank's RedHot sauce
150g butter

METHOD

First, make the sauce by melting the butter in a pan and adding Frank's sauce and the JFC seasoning.

If you like your wings really hot, add a hit of cayenne or some crazy hot pepper sauce to your pan too.

Now add around an inch of oil to a heavy-bottomed pan and heat to 180°C.

Fry the wings until they start to brown – around ten to fifteen minutes.

Remove the wings and pat dry.

Toss the wings in the sauce in a large bowl until evenly coated.

Serve with a blue cheese dip.

POPCORN SHRIMP

I FIRST ATE THIS AT THE WINKING PRAWN DOWN IN DARTMOUTH. THEY SERVE THE FRESHEST OF FRESH SHRIMP, QUICKLY FRIED UP IN A BEAUTIFULLY SEASONED BATTER AND THEN PILED HIGH IN A SUB ROLL. I RECOMMEND A TRIP DOWN THERE, NOT ONLY FOR THE POPCORN SHRIMP, BUT ALSO THE BEAUTIFUL VIEW OF THE BAY AS YOU EAT.

WASH IT DOWN WITH A GOOD DRY IPA.

INGREDIENTS

Allow 100–150g raw prawns per person
Bowl of tempura or beer batter (see page 140)
Soft white sub rolls
Vegetable or sunflower oil

For the JFC seasoning (It's like KFC's but it's my version)
⅔ tbsp salt
⅓ tbsp dried oregano leaves
1 tbsp celery salt
1 tbsp ground black pepper
1 tbsp dried mustard
4 tbsp paprika
2 tbsp garlic salt
1 tbsp ground ginger

For the seaside sauce, mix together
4 tbsp ketchup
4 tbsp mayo
4 splashes Worcestershire sauce
4 splashes tabasco
Juice of ½ lemon

SERVES 4-6

METHOD

Wash the prawns, pat dry and pop into a mixing bowl.

Add 1–2 tbsp of the JFC seasoning and mix well.

Add the batter to the seasoned prawns and stir gently to coat well.

Heat 5–6 inches of vegetable or sunflower oil in a deep pan until it reaches 180°C.

In batches fry the prawns for around three minutes until golden, and drain on kitchen roll.

Slit the sub roll lengthways, add the prawns and cover liberally in the seaside sauce.

PHILLY CHEESESTEAK

I ABSOLUTELY ADORE A PHILLY CHEESESTEAK. THIS ONE IS EVEN BETTER AS WE'RE USING THE BEER CHEESE NACHO DIP TO GIVE AN EXTRA LAYER OF GOOEY, CHEESY FUN.

THE COOK'S TRICK IN TAKING THIS FROM BEING A GOOD PHILLY CHEESESTEAK TO BEING AN EPIC ONE IS TO COVER THE STEAK AND CHEESE AS THEY COOK WITH A SMALL BOWL OR PAN LID SO IT ALL STEAMS AND MELTS. THEN SCOOP THE WHOLE THING INTO YOUR SANDWICH.

INGREDIENTS

2 large hot dog/sub rolls, split open lengthways most of the way
200g bavette steak, chilled
½ an onion, finely minced
½ a green pepper, finely minced
6 provolone cheese slices
Beer cheese nacho dip (see page 146)
Salt and pepper

METHOD

Gently warm some of the beer cheese nacho dip and spread a generous layer inside each split sub roll.

Remove steak from the fridge and slice razor thin.

Season the steak with a pinch of salt and pepper.

Fry the onion and pepper in a little oil until they soften.

Add the sliced steak and flash-cook for a further two to three minutes.

Scoop everything together into two neat lengthway piles in the middle of the pan – the right size to fit in your roll.

Layer three cheese slices over each pile in the pan then cover for thirty seconds with the lid or a bowl.

Gently scoop one of the meaty, cheesy piles of gorgeousness into each roll.

Open your mouth, insert one end of the sandwich, and keep pushing until it has gone.

WELSH RAREBIT

I'VE ALWAYS WONDERED WHY WELSH RAREBIT WAS CALLED RAREBIT. YOU'LL BE PLEASED TO KNOW I GOOGLED IT SO YOU DON'T HAVE TO. AS WE ALL PROBABLY GUESSED, IT IS A CORRUPTION OF THE WORD 'RABBIT' AND IS ESSENTIALLY FANCY CHEESE ON TOAST, BUT IT ACTUALLY NEVER WENT ANYWHERE NEAR A RABBIT. WEIRD.

THE QUESTION REMAINS: WHY THE HELL WAS IT CALLED 'RABBIT' IN THE FIRST PLACE? AH WELL – ON WE GO TO HOW TO MAKE IT.

INGREDIENTS

250g grated, strong Cheddar
1 tbsp butter
2 tsp Worcestershire sauce
1 level tsp dry mustard
2 tsp flour
Pinch of black pepper
60ml dark, malty beer or more as needed
4 slices of bread

METHOD

Add all the ingredients apart from the bread to a pan.

Heat gently and stir to combine, adding just enough beer to get to the desired consistency. We all love beer but nobody is going to be very pleased with a soggy, limp rarebit. A thick paste is ideal.

Place the bread slices under the grill, toasting just one side (to make a bit more luxurious, butter this side before you grill it).

Flip the bread over, smear the paste over the untoasted side and slide back under the heat until brown.

That's it – job done. Easy peasy.

SERVES 4

POT-BRAISED CHICKEN THIGHS

FAMILY FAVOURITE IN OUR HOUSE IN PLACE OF THE REGULAR SUNDAY ROAST.

I LIKE TO BUY A WHOLE CHICKEN AND CHOP IT INTO PIECES MYSELF. MUCH CHEAPER THAT WAY AND IT MAKES YOU FEEL MANLY. SEARCH YOUTUBE FOR INSTRUCTIONS ON THAT.

INGREDIENTS

1 whole chicken or 8 chicken thighs
Flour
Salt & pepper
Olive oil
1 onion – coarsely sliced
4 carrots – peeled and halved
2 sticks of celery, roughly chopped
1 fennel bulb, sliced (optional)
250ml beer
250ml chicken stock
Fresh thyme

METHOD

Season the chicken with a generous pinch of salt and pepper then dredge in flour. Shake off the excess.

Heat 3 tbsp olive oil in a deep, ovenproof dish.

Add chicken pieces and cook until golden brown all over – around ten minutes.

Remove chicken pieces and set aside.

Add veg and sauté until soft.

Add the beer, stock and thyme and bring to a boil.

Reduce to a gentle simmer and add the chicken thighs, skin side up. The underside of the chicken should be submerged with the skin sitting above the water line.

Cook gently on the hob or uncovered in an oven (150°C) until the chicken reaches an internal temperature of 75°C.

Use the cooking liquid as gravy and serve with a big pile of mashed potatoes.

STEAK & ALE PIE

THIS MUST BE ONE OF THE MOST POPULAR AND TASTIEST WAYS TO COOK WITH BEER. AN ABSOLUTE CLASSIC. MAKE IT EPIC BY PURCHASING REALLY GOOD BEEF FROM YOUR LOCAL BUTCHER. ALSO, AVOID HOPPY BEERS AS THE BITTERNESS WILL COME THROUGH IN THE DISH. MANY YEARS AGO I RUINED A PERFECTLY GOOD BEEF STEW BY USING A VERY HOPPY IPA AND THE BITTERNESS JUST INTENSIFIED. I RUINED MY STEW SO YOU DON'T HAVE TO.

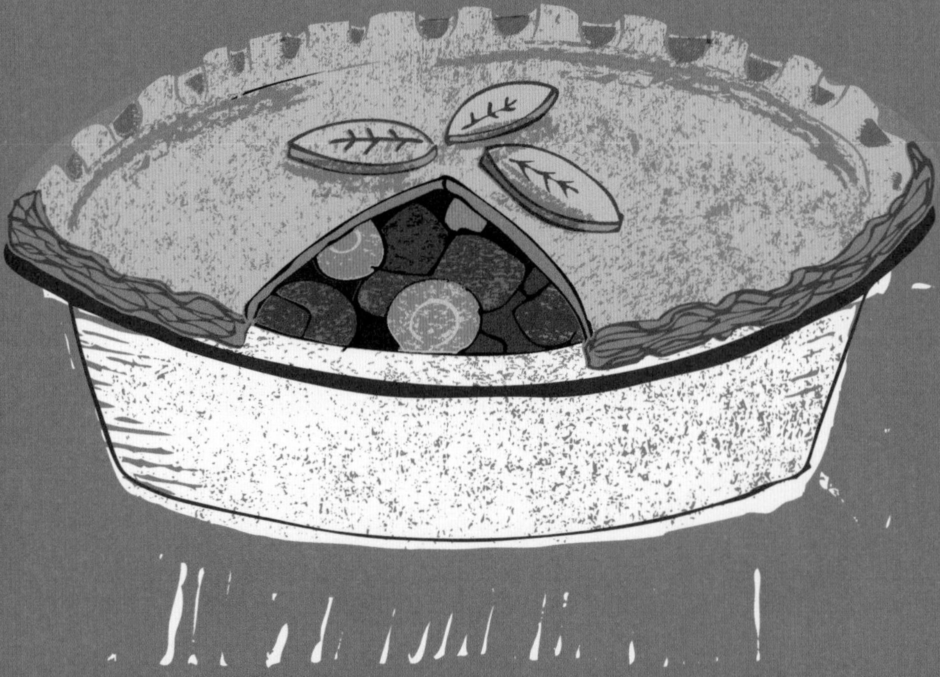

INGREDIENTS

The stew

1kg stewing steak/beef shin – chopped into 1" cubes
Plain flour
Salt & pepper
1 onion, diced
2 sticks celery, coarsely chopped
6 large carrots, coarsely chopped
1 tbsp tomato puree
1 bay leaf
500ml good beef stock
500ml either malty dark ale or stout

The pastry

650g plain flour, plus extra for dusting
250g lard or cold butter, diced, plus extra for greasing
1 egg, beaten

METHOD

First make the pastry.

Crumble the flour and lard, or butter, together with a generous pinch of sea salt until completely combined. Add around 200ml ice-cold water to make a soft dough.

Knead to combine then wrap in clingfilm and leave to rest in the fridge for at least one hour.

Generously season the meat with salt and pepper.

Tip flour into a bowl and dredge the beef cubes so they get a light coating.

In two to three batches, brown the beef cubes in a deep casserole pan over a medium heat, transferring the cooked batches to a bowl.

When all the beef has browned, add the onions, carrots and celery to the pan. Cook until softened and starting to brown.

Then add the tomato puree and cook for a further minute or two.

Add the stock and beer. Scrape all the sides and bottom of the pan to get the dark brown bits mixed in – this is flavour!

Add the beef back to the pan and add the bay leaf.

Lower the heat to the gentlest of simmers (one bubble every twenty seconds) and allow to cook, lid on (alternatively, place the cooking pot in an oven at 150°C for the same amount of time).

Check every hour or so and add more stock if it begins to dry out.

Cook for three to four hours and then test the beef – it should be soft enough to cut with a blunt spoon.

Tip the mixture into a pie dish and top with the pastry, crimping down the edges to make a good seal.

Brush the pastry with the beaten egg.

Slash the lid to allow steam to escape.

Bake in a hot oven (180°C) for around twenty-five minutes until the pastry is golden brown.

FUN
STUFF

KEGGING BEER

Why would you not want your own fresh craft beer on tap in your own home? Of course you would. Everyone would.

Kegging beer not only makes it easy to dispense, it saves you time and a bit of hassle as I find bottling beer a faff.

Once the beer has completed its fermentation and you have transferred it to a keg you have a number of choices as to how to set it up.

The simplest is to attach what's called a party tap – a short length of beer line with a simple press-to-open tap on the end. With a little extra effort, though, it's really easy to mount some proper bar fonts on a bit of wood and then mount that wood somewhere convenient.

In my shed at home, I have a big cylinder of CO2 gas that is connected to three beer kegs. I drilled holes and then mounted three beer taps on the front of my shed door (much to my wife's delight). Behind the door, the kegs are connected to the back of the taps. It's the perfect shed. A brew-shed if you will. Perfect for summer garden parties. Though not so perfect if you have small children who pull the tap handle down and watch 20L of your finest home brew dispense itself all over the patio. Yep – still not forgiven him for that five years later . . .

I'm pretty chuffed with my brew-shed, but there are those who go one step further and convert used fridges into a dispensing system. It makes perfect sense. A small under-the-counter fridge is just the right size to house a 20L keg of beer and a small gas cylinder. Simply drill through the door, mount your tap on the outside, connect it all up and fire up the fridge. A portable chilled beer dispensing system in your own home.

A CHAP I KNOW, LUKE WELSH ADAMS (BREWJUNKY), BUILT A REALLY COOL SET-UP. INSTEAD OF USING A FRIDGE HE CONSTRUCTED WHAT'S ESSENTIALLY A BOX ON WHEELS WITH TWO COMPARTMENTS. ONE SIDE IS INSULATED AND HOLDS THE BEER KEG, THE OTHER HOLDS THE GAS. ON TOP HE'S MOUNTED A BEER TAP.

IF YOU ARE SO MINDED, I'VE INCLUDED HIS BASIC PLANS BELOW, AND HERE ARE THE MATERIALS YOU'LL NEED:

KEGERATOR KIT
19L CORNELIUS KEG
CO2 CYLINDER + REGULATOR
3/8" CO2 & FLUID PIPES
DISPENSING TAP

MATERIALS USED – APPROX.
8m × C16 47 × 125 FRAMING TIMBER
MDF SHEET – 9 × 606 × 1220mm
8 × 90-DEGREE STEEL BRACKETS
75mm DECK SCREWS
PVC MEMBRANE
2 × HINGES
STEEL HANDLE
CHECKER PLATE EFFECT SELF-ADHESIVE VINYL
DRIP TRAY
DECK BOARD OFFCUT (TO MAKE THE TAP TOWER)
10 × FEATHER-EDGE FENCE BOARD 100 × 11mm × 1.8m
3 × CELOTEX 50mm INSULATION 450 × 1200mm

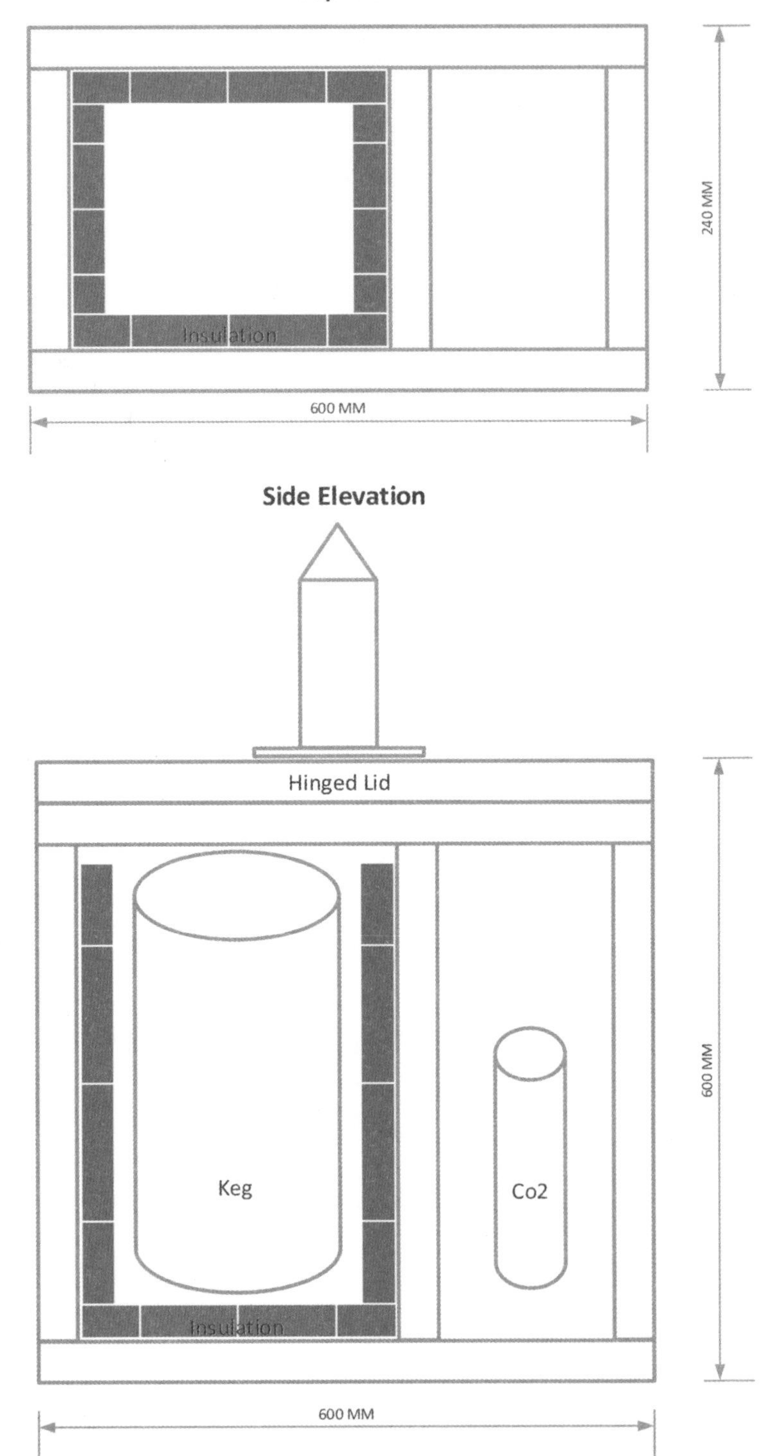
Top View
Insulation
240 MM
600 MM
Side Elevation
Hinged Lid
Keg
Co2
Insulation
600 MM
600 MM

Craft
Ale
BrewJunky

BREWING RESOURCES

INGREDIENTS & KIT

For years my 'go-to' for all brewing equipment and ingredients has been the Malt Miller. Rob and his crew provide a huge range of the freshest ingredients at great prices as well as every last bit of kit you could possibly ever need. They also stock a good range of the ever-so-important chemicals for keeping everything nice and clean. Speedy service too. Tell him I sent you.

www.themaltmiller.co.uk

ONLINE

Basic Brewing Radio – I've listened to James Spencer's podcasts for several years. They've taught me everything I know about brewing as well as inspired me to brew more! It was a big moment for me when James agreed to write the foreword to this book. He's my home-brew hero!

www.basicbrewing.com, or search iTunes podcasts

Beersmith – excellent home-brewing software to keep track of all your brewdays as well as build your own recipes. It also has an online 'cloud' facility to share recipes with others and can be run from your iPad or iPhone.

http://beersmith.com

FOOD

If you enjoyed this book then please jump on Amazon and order my other book – *Grillstock, The BBQ Book*, and learn how to BBQ like a bad-ass.

HOME-BREW BOOKS

Brewing Classic Styles – Jamil Zainasheff and John Palmer run through eighty classic styles of beer and give example recipes for how to brew each one.

Brew Your Own British Real Ale – A solid practical guide by Graham Wheeler covering methods and equipment, and giving around a hundred recipes.

How To Brew – John Palmer's guide to everything you need to know to brew beer – one for when you have mastered the basics.

Clone Brews – Tess and Mark Szamatulski offer up recipes for you to make your own versions of over two hundred commercial beers, covering virtually every beer style under the planet.

Radical Brewing – Randy Mosher delves into crazy, exotic and sometimes wacky beers, beer history and beer ingredients. Great fun.

The Complete Joy of Homebrewing – Charlie Papazian's book is generally regarded as the original home-brew bible. My copy is well thumbed.

Designing Great Beers – Ray Daniels breaks down in micro detail the art of building beer recipes – this is one for when you want to start getting real geeky on brewing.

Brew – *Great British Bake Off* finalist James Morton loves his brewing as much as his baking. This is a very accessible, beautifully made book that goes into the brewing processes in a little more depth than this book. A great next step without being overwhelming.

***Brew Your Own* Magazine** – US-based and a little pricey to import, but luckily you can download it on the iPad!

INDEX

ABOUT THE AUTHOR

In 2010 I co-founded Grillstock, the biggest BBQ & music festival outside of the US and a chain of smokehouses. In 2015 I wrote *Grillstock: The BBQ Book*, which went on to be the bible for smoking and grilling meat.

BBQ is a huge passion of mine – and beer another. What's better than sipping on cold, fresh beer and grilling some tasty meat? We've been doing it for thousands of years!

Back in 1997, I received a home-brew kit for Father's Day from my newborn son (with help from his mum). I made the beer and, as with pretty much anyone's first batch of home brew, it sucked. It was cloudy, smelly and flat, and I had to pour it down the drain.

I decided to give it another shot, and the next batch was actually okay. Not great, but it resembled beer enough for me to serve to some friends at a BBQ. To my surprise, one or two of them even went back for refills. That's when I got the bug and became hooked on this wonderful hobby of home brewing. In how many other pastimes can you invest a couple of hours doing something you really enjoy and then get forty pints of beer afterwards as a bonus?

Fast forward a few years and my tiny little shed that houses my beer kegs and home-brew gear usually has at least three different craft beers on tap. It's affectionately known in the village as 'The Crow', and my neighbours know they are welcome to drop in to pull themselves a pint or two even if I'm not at home.

Like a good BBQ, beer brings people together and I think that's why I love it so much.

Welcome to your new hobby.

ACKNOWLEDGEMENTS

I have a lot of people to thank, so here goes.

Marie-Louise, Noah and Jake for your unswerving confidence in me and also for not getting cross when our house smells like a brewery.

My compadre Ben, for being my partner-in-crime all these years and your insanely good graphic design talent.

Mum, Dad, Bob, Brett, Beth, Owain and the rest of my extended family – love you loads.

Harriet for keeping the beer paunch at bay.

The Chalford crew for drinking my beer.

Ruth for all those fabulous illustrations!

Adam Strange for taking another punt on me. Apologies for all your new, naughty hobbies.

Nithya and the rest of the Little, Brown crew for keeping me on the straight and narrow.

Emily at Charles Faram for your wonderful insight and help on all things hoppy.

Rob @ Malt Miller for supplying awesome ingredients.

Richard @ Stroud Brewery for letting me 'help' on a brewday.

James from Basic Brewing Radio for teaching me pretty much everything I know about brewing beer!

All the brewers who have generously given their time and recipes to help make this book, namely: Ian & Sally @ Brick Brewery, Niall & Gazz @ Tiny Rebel Brewery, Andrew @ Wild Card Brewery, Andrew & Brett @ Wild Beer Brewery, Justin @ Moor Beer, Greg @ Stroud Brewery, Jasper @ Camden Brewery, Eddie @ Harbour Brewery, Jiri @ Budvar and Garrett @ Brooklyn Brewery.

SPHERE

First published in Great Britain in 2017 by Sphere

The authors and publishers would like to thank the following breweries for contributing interviews, photographs and recipes to the book: Brick Brewery, Tiny Rebel Brewery, Wild Card Brewery, Wild Beer Brewery, Moor Beer, Stroud Brewery, Camden Brewery, Harbour Brewery, Budvar and the Brooklyn Brewery.

10 9 8 7 6 5 4 3 2 1

A CIP catalogue record for this book is available from the British Library.

ISBN 978-0-7515-6937-7

Printed and bound in China

Papers used by Sphere are from well-managed forests and other responsible sources.

Sphere
An imprint of
Little, Brown Book Group
Carmelite House
50 Victoria Embankment
London EC4Y 0DZ

An Hachette UK Company
www.hachette.co.uk

www.littlebrown.co.uk